Earth Resources

a dictionary of terms and concepts

Arrow Reference Series

eral Editor: Chris Cook

Earth Resources

a dictionary of terms and concepts

David Dineley
Donald Hawkes
Paul Hancock and
Brian Williams

Arrow Books

Arrow Books Limited
3 Fitzroy Square, London W1

An imprint of the Hutchinson Publishing Group

London Melbourne Sydney Auckland
Wellington Johannesburg and agencies
throughout the world

First published Arrow Books 1976
© David Dineley, Donald Hawkes, Paul Hancock,
and Brian Williams 1976

Set in Monotype Times
Made and printed in Great Britain
by The Anchor Press Ltd
Tiptree, Essex

ISBN 0 09 913690 2

Using this Dictionary

Not all the terms defined in this dictionary are cross-referenced to one another every time they occur. Only when the understanding of a term used in an entry adds to the student's comprehension of the particular area under discussion will the term be marked in small capitals, thus:

Acidization. The acid treatment of wells drilled into LIMESTONE AQUIFERS, for which hydrochloric acid is usually employed.

A single arrow has been used for 'See'; double arrows for 'See also'.

A list of books for further reading can be found at the back of the book.

Table of Elements

Element	Symbol	Atomic weight*	Terrestrial abundance (ppm)	Year discovered
Actinium	Ac	(227)‡	3×10^{-10}	1899
Aluminium	Al	26·9815	81 000	1825
Americum	Am	(243)		1944
Antimony	Ab	121·75	1	BC
Argon	Ar	39·948	0·04	1894
Arsenic	As	74·9216	5	1649
Astatine	At	(210)		1940
Barium	Ba	137·34	250	1808
Berkelium	Bk	(249)		1949
Beryllium	Be	9·0122	6	1797
Bismuth	Bi	208·980	0·2	c.1739
Boron	B	10·811	3	1808
Bromine	Br	79·909	1·62	1826
Cadmium	Cd	112·40	0·15	1817
Calcium	Ca	40·08	36300	1808
Californium	Cf	(251)		1950
Carbon	C	12·01115	320	BC
Cerium	Ce	140·12	46·1	1803
Cesium	Cs	132·905	7	1860
Chlorine	Cl	35·453	314	1774
Cobalt	Co	58·9332	23	1742
Copper	Cu	63·54	70	BC
Curium	Cm	(247)		1944
Dysprosium	Dy	176·26	4·47	1886
Einsteinium	Es	151·96		1955
Erbium	Er	(253)	2·27	1843
Europium	Eu	18·9984	1·06	1901
Fermium	Fm	(223)		1955
Fluorine	F	157·25	900	1771
Francium	Fr	(223)		1939
Gadolinium	Gd	157·25	6·36	1880

Table of Elements

Element	Symbol	Atomic weight*	Terrestrial abundance (ppm)	Year discovered
Gallium	Ga	69·72	15	1875
Germanium	Ge	72·59	7	1886
Gold	Au	196·967	0·005	BC
Hafnium	Hf	178·49	4·5	1923
Helinium	He	4·0026	0·003	1895
Holmium	Ho	164·930	1·15	1878
Hydrogen	H	1·00797	1300	1766
Indium	In	114·82	0·1	1863
Iodine	I	126·9044	0·3	1811
Iridium	Ir	192·2	0·001	1804
Iron	Fe	55·847	50000	BC
Krypton	K	83·80	98×10^{-6}	1898
Lanthanum	La	138·91	18·3	1839
Lawrencium	Lv	(257)		1961
Lead	Pb	207·19	16	BC
Lithium	Li	6·939	65	1817
Lutetium	Lu	174·97	0·75	1907
Magnesium	Mg	24·312	20900	1755
Manganese	Ma	54·938	1000	1774
Mendelevium	Ml	(258)		1955
Mercury	Hg	200·59	0·5	BC
Molybdenum	Mo	95·94	15	1782
Neodymium	Nd	144·24	23·9	1885
Neon	Ne	20·183	7×10^{-5}	1898
Neptunium	Np	(237)		1940
Nickel	Ni	58·71	80	1751
Niobium	Nb	92·906	24	1801
Nitrogen	N	14·0067	46·3	1772
Nobelium	No	(253)		1952
Osmium	Os	190·2	0·005	1804
Oxygen	O	15·9994	466000	1772
Palladium	Pa	106·4	0·010	1803
Phosphorus	P	30·9738	1180	1669
Platinum	Pt	159·09	0·005	1735
Plutonium	Pl	(242)		1940
Polonium	Po	(210)	3×10^{-10}	1898
Potassium	K	39·102	25900	1807
Praesodymium	Pr	140·907	5·53	1885

Element	Symbol	Atomic weight*	Terrestrial abundance (ppm)	Year discovered
Prometheum	Pm	(145)		1947
Protactinium	Pa	(231)	8×10^{-7}	1917
Radium	Ra	(226)	13×10^{-6}	1898
Radon	Rn	(222)		1900
Rhenium	Re	186·2	0·001	1925
Rhodium	Rh	102·905	0·001	1803
Rubidium	Rb	85·47	310	1861
Ruthenium	Ru	101·07	0·004	1844
Samarium	Sm	150·35	6·47	1879
Scandium	Sc	44·956	5	1879
Selemium	Se	78·96	0·09	1818
Silicon	Si	20·056	277200	1823
Silver	Ag	107·870	0·10	BC
Sodium	Na	22·9898	28300	1807
Strontium	St	87·62	300	1787
Sulphur	S	32·064	520	BC
Tantalum	Ta	180·948	2·1	1802
Technetium	Tc	(99)		1937
Tellurium	Te	127·60	0·002	1798
Terbium	Tb	158·924	0·91	1843
Thallium	Tl	204·37	3	1861
Thorium	Th	232·038	11·5	1828
Thulium	Tm	168·934	0·20	1879
Tin	Sn	118·69	40	BC
Titanium	Ti	47·90	4400	1791
Tungsten	W	183·85	69	1783
Uranium	U	238·03	4	1789
Vanadium	V	50·942	150	1830
Xenon	X	131·30	1×10^{-6}	1898
Ytterbium	Yb	173·04	2·66	1907
Yttrium	Y	88·905	28·1	1794
Zinc	Zn	65·37	132	Before 1400
Zirconium	Zr	91·22	220	1789

Names in italics appear as separate entries in the text.

* Based on the weight of carbon–12 as the standard.

† Parts per million = grams per metric tonne (most of data after K. Rankama and T. G. Sahama, *Geochemistry* (1950).

‡ Figure in brackets is mass number either of isotope with longest half-life or a better-known one.

(Based on *Encyclopaedia Britannica*)

Geological Time-Scale

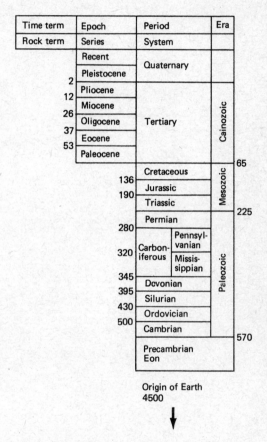

Time term	Epoch	Period	Era
Rock term	Series	System	
	Recent	Quaternary	
	Pleistocene		
2	Pliocene		Cainozoic
12	Miocene		
26	Oligocene	Tertiary	
37	Eocene		
53	Paleocene		
65		Cretaceous	Mesozoic
136		Jurassic	
190		Triassic	
225		Permian	Paleozoic
280		Carboniferous — Pennsylvanian	
320		Carboniferous — Mississippian	
345		Devonian	
395		Silurian	
430		Ordovician	
500		Cambrian	
570		Precambrian Eon	

Origin of Earth
4500

↓

Figures refer to millions of years

Introduction

The aim in producing this small volume has been to provide a compact and not too technical source of information about the many natural commodities upon which our industrial society depends. Not only are the resources themselves listed, but the means of discovering, assessing and retrieving them are briefly dealt with. While metallic ores, building materials and chemical deposits are obvious and important resources in any country, the hidden – or at least commonly unseen – sources of wealth such as oil and gas have been much in the news for several years now. As the 'energy crisis' continues, our interest in these resources will certainly not diminish, but there is in the essential resource, water, an importance no less than that of any other national asset. Rich and poor nations alike are anxious to raise or maintain living standards and it is no exaggeration to say that this means an intensified search for supplies of potable ground water and a better management of water resources. Water is the one great reusable natural resource we have. Its importance, we hope, is reflected in the treatment this topic receives in this dictionary.

The literature on resources, environmental spoliation, conservation, pollution and on resource technology is now vast and continues to grow each year. A few selected references of the more easily obtainable kind are given in the main body of the book; a further selection is given at the end. Principal sources of statistical and other more technical information are listed below. The special reports issued by *The Times* newspaper are worth special mention as excellent and up-to-date sources of data.

Under each entry certain words may be in small capitals: these have an entry in the main text. Immediately below are two entries which are at the same time all-important and yet the most general of all. The study of the planet earth is geology, that of its oceans oceanography, or oceanology as it has now become. These sciences are now of national importance to the great powers and indeed to most countries of the world. Several excellent dictionaries and encyclopedias of geology and oceanography are listed.

In writing about the origin, character and technology of natural resources most of the words used are those of existing sciences or technologies. There are, however, certain terms used by the economist,

13

the media and the man in the street in discussing 'resources', and the usage may differ between one speaker or writer and the next. It is most important that these indispensable terms be understood and properly employed. A commodity may be valued for its usefulness or its rarity; it may be easy and cheap to extract or utilize or it may be subject to wide fluctuations in market value. To understand why this may be so involves more than simple mineralogy or (less simple?) economics and we hope that the few definitions in this general area will be found useful.

Finally, perhaps, it should be emphasized that this is not a dictionary of geology. Some geological terms do occur in the following pages, but they are considered in the context of earth resources.

The earth. One of the inner group of four small planets of the solar system. It has an equatorial diameter of 12 756·8 kilometres and a mass of $6·5 \times 10^{21}$ tonnes. It possesses a gaseous envelope (atmosphere) and is partially covered by water (hydrosphere). Some 88 different chemical elements are present. The composition of the outer solid part (the crust) is estimated from rock analyses. The nature of the interior is suggested from a study of volcanoes and volcanic material from deep within the earth and from the records of earthquake shocks which move through and round the planet. Apparently the distribution of the chemical elements must be very uneven throughout the planet because the nature of the matter present changes with depth. The earth is thought to have a concentric structure as illustrated in Figure 1.

Overall, the bulk of the earth is made up of ten elements, the proportions varying from depth to depth. Even in so inhomogeneous a planet the composition of the crust is remarkably variable. A mere 0·375 per cent of the earth's mass, the crust varies broadly in composition between the oceanic and the continental areas, and within the continents the elements are spread very irregularly.

The crust is thickest beneath the continents where its composition is dominated by the elements silicon, aluminium and oxygen. Beneath the oceans, however, magnesium and iron also constitute much of the crust. The crust is composed of minerals, silicates being by far the greater proportion. Nine elements seem to make up the bulk of its composition (see Figure 1). The crust is the source of virtually all mineral resources.

The mantle is denser than the crust and may be more uniform around the earth, though it probably becomes denser with depth. A few very dense magnesium and iron-rich crystalline rocks are thought to originate from the mantle but it does not otherwise reach the surface. (Figure 2.)

The atmosphere is composed of nitrogen, oxygen, water vapour and argon. It is continuously moving and thus is essentially uniform in composition. Less than 0·1 per cent is composed of other gases. A

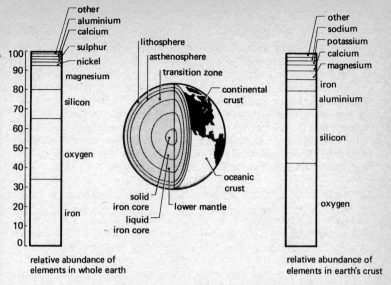

Figure 1. The concentric structure of the earth and the relative abundances of elements within the earth. (After Press and Siever.)

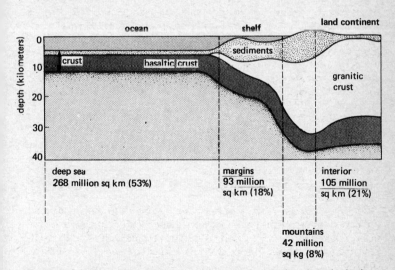

Figure 2. Cross-section of the crust of the earth showing the principal rock types and the areas they occupy (in millions of square kilometres and percentages). (After Laporte.)

15

virtually limitless source of oxygen and nitrogen, it has played an important role in the evolution of fossil fuels, carbonate rocks and the weathering of most rocks and minerals throughout geological time.

The hydrosphere includes all the water on or near the surface of the earth. Its role is vital not only for life but also in the evolution of most kinds of rocks. Water covers over 70 per cent of the earth's surface and comes as near as any substance to being the universal solvent. (Figure 3.)

The geological cycle is the unending movement of materials in and above the crust of the earth that is perhaps one of the most distinguishing features of our planet. Once the primeval earth had acquired an

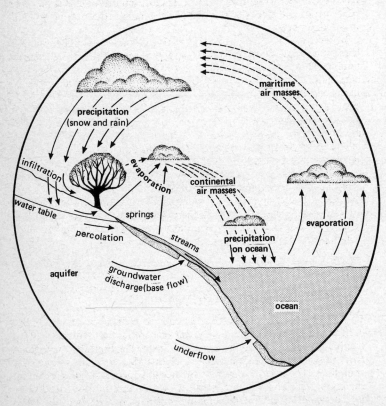

Figure 3. The hydrological cycle. The pathways of movement in the hydrosphere. (After Brown *et al.*, 1972.)

atmosphere and a hydrosphere, chemical reactions between them must have begun. Volcanic activity brought new matter to the earth's surface to form volcanic (or igneous) rock. This was weathered and decomposed, and the products fell, were washed or simply blown away to settle elsewhere as a sediment. Such sediments accumulated locally in the seas and oceans, forming great masses here and there and consolidating into stratified rocks. When earth movements took place these strata were deeply buried, affected by the heat and pressures within the earth, attacked by gases and fluids from within the crust. Ultimately they may have been remelted and intruded into the outer crust or erupted as volcanic matter once again. Such a cycle may take many millions of years, and it has no doubt evolved and changed as the earth has grown older. (Figure 4.) One of its most important effects has been to influence and intensify the highly irregular distribution of minerals in the outer crust.

In recent years the concept of 'plate tectonics' has influenced geologists' thinking about the nature of the geological cycle and its bearing upon natural resources. This concept holds that the crust of the earth consists of several thin but rather rigid 'lithospheric plates' which are in process of formation and destruction along their various margins. The surfaces of these plates thus appear to move as geological time passes. They rest upon a plastic layer of the mantle known as the asthenosphere, some 60 or more kilometres down. Upon the surfaces of the plates rest the continents, and from time to time some of the continents are brought into collision by the movement of the plates, and zones of intense rock movement and deformation known as orogenic belts are produced. The collisions in fact occur at the destructive margins of the plates, over which there may also be situated ocean trenches and volcanic arcs. Thus the crust of the earth is not static but is dynamic and has been in continuous movement during the 4500 million years of the earth's existence (Figure 5).

The Story of the Earth, Institute of Geological Sciences, London, 1973.
L. F. Laporte, *Encounter with the Earth*, Harper & Row, 1975.
H. W. Mennard, *Geology Resources and Society*, W. H. Freeman, 1974.
F. Press and R. Siever, *Earth*, W. H. Freeman, 1974.

Ocean. The oceans of the world occupy about 70 per cent of the earth's surface and overlie the oceanic basins of basaltic parts of the crust, as distinct from the continental or granitic crust. Geologically they extend from the edges of the continental shelves rather than the present-day shorelines. The average depth of the ocean is nearly 4 kilometres, but there are deeps and 'mountain' ridges of great size locally. (See Figure 6.) The water in the oceans contains enormous quantities of dissolved mineral matter and virtually all the naturally occurring elements. The ocean sediments include wind- and stream-borne detritus, biogenic

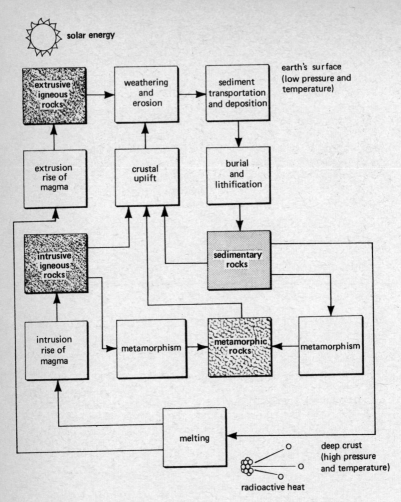

Figure 4. The geological cycle showing the rocks and the processes which modify and form them in the crust of the earth. The cycle is kept moving by the energy from the sun and from radioactive decay within the earth itself. (After Laporte.)

18

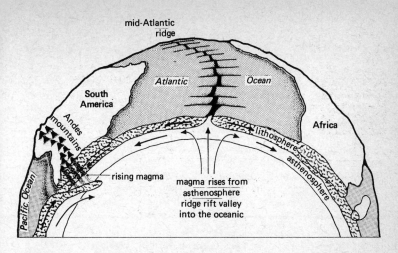

Figure 5. Plate tectonics. The current theory which holds that the rise of magma beneath oceanic ridges causes the movement of relatively rigid plates of the outer crust and brings about the movement of continents and the formation of the great earthquake and volcanic mountain regions. (After Wiley.)

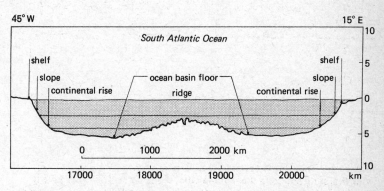

Figure 6. A 'typical' ocean basin is a broadly symmetrical structure with a ridge in its mid-part. (After Wiley.)

19

materials such as calcite, silica, phosphate and other organic substances, and nodules of manganese and other metals. Attempts to mine these nodules have been delayed by technological difficulties.

The average concentrations of the seven most abundant solutes in seawater are, in milligrammes per litre, sodium, 1 050 000; chlorine, 19 000; magnesium, 1350; sulphur, 885; calcium, 400; potassium, 380; bromine, 65. These elements are extracted commercially by various processes.

Bromine is principally derived from seawater, and magnesium is also extracted in considerable volumes. The extraction of other elements has occupied the attentions of many workers and an impetus to this has been provided by the discovery of deep hollows or 'pools' of hot, highly saline brine carrying abnormally high concentrations of barium, copper, iron, zinc and other metals in the Red Sea.

K. K. Turekian, *Oceans*, Prentice Hall, 1968.
R. Revelle, The Ocean, *Scientific American*, vol. 221 (1969), pp. 54–65.
E. Wenke, Jr, The Physical Resources of the Ocean, ibid., pp. 166–76.
W. Bascom, Technology and the Ocean, ibid., pp. 199–217.
W. S. Wooster, The Ocean and Man, ibid., pp. 218–34.

A

Abrasives. After relatively simple processing many minerals or rocks can be employed as abrasives. Most manufactured abrasives are also derived from rocks or minerals. Not all abrasives are particularly hard; for some purposes soft abrasives are required. Grains to be used as abrasives should break down only slowly, and they should develop fresh cutting surfaces when being worn. Abrasive materials for bonding should be capable of withstanding high temperatures. Other abrasives are used in powder form, coated on a backing of paper or cloth, or employed in block form. Most of the rocks and minerals used as abrasives are described elsewhere in this book, and they include diamond, corundum, emery, garnet, silica, sandstone, basalt, feldspar, granite, schist, perlite, pumice, calcite, chalk, clay, diatomite, dolomite, limestone, talc, tripoli, bentonite and china clay.

Absorption on minerals. Minerals which absorb material assimilate or incorporate that material, in contrast to those which adsorb material on the surfaces of particles. Some clays are able to absorb infra-red radiation. DIATOMITE is a rock used for absorption. (◊ADSORPTION ON MINERALS.)

Abstraction well. ◊PRODUCTION WELL.

Acid-activated clays. ◊ADSORBENT AND BLEACHING CLAYS.

Acidity. Acidity in water is the opposite property of ALKALINITY, and is the ability of water to neutralize a base or alkali added to it. Fortunately it is uncommon to find free mineral acids in ground waters. Examples are found in coalmining areas where acid mine drainage is encountered. These acid waters originating from surface or underground mine workings are almost always sulphuric in type resulting from the breakdown of metallic sulphides. (Waters with a pH value below 7 are called acid-type waters but free mineral acidity is only present when the pH is below 4·5.) (◊pH.)

Acidization. The acid treatment of wells drilled into LIMESTONE AQUIFERS, for which hydrochloric acid is usually employed. The introduction of acid cleans the walls of the well and enlarges solution channels, thereby possibly improving the well yield by a few hundred per cent.

Acid rocks. ⟡IGNEOUS ROCKS.

Actinolite. ⟡AMPHIBOLE; ASBESTOS.

Activated bauxite. ⟡ADSORBENT AND BLEACHING CLAYS.

Adobe. ⟡BRICK CLAYS.

Adsorbent and bleaching clays. Clays capable of chemically adsorbing oils, insecticides, alkaloids, vitamins, carbohydrates and other materials, either after being activated or in their natural state (⟡ADSORPTION ON MINERALS). Bleaching clay, or bleaching earth, can adsorb colouring matter in oils. They are used for refining and decolorizing mineral and vegetable oils.

The three principal materials used for adsorption or bleaching are Fuller's Earth, acid-activated clays and activated bauxite.

Fuller's Earth is a naturally active clay rich in the CLAY MINERAL montmorillonite. Both Fuller's Earth and those clays which are active after treatment belong to the category of BENTONITES. The best bentonites for use in bleaching or adsorption are of the non-swelling type. Some bentonitic adsorbent clays also contain a significant proportion of the clay mineral attapulgite. Kaolin, another clay mineral, has been used for bleaching some oils.

Important deposits of Fuller's Earth occur in the USA, England and Japan, and there are clays which are of value after acid-activation in many countries of the world.

Adsorption on minerals. Minerals capable of adsorbing liquids or gases retain that material by its adhesion as very thin layers on the surfaces of the mineral particles. Clay minerals in the BENTONITE group are, for example, able to adsorb, in their natural state or after acid-activation, oils and colouring matter. They are used as ADSORBENT AND BLEACHING CLAYS.

Aeration zone. Water infiltrates downwards from the ground surface through the soil and rocks to a level where all the VOIDS and FRACTURES are filled with ground water. The zone where the openings (or voids) in the rocks are only partially filled with water is termed the zone of aeration. The zone of aeration comprises three subzones – the SOIL WATER ZONE, the INTERMEDIATE ZONE and the CAPILLARY FRINGE (Figure 7). Immediately below the zone of aeration is the ZONE OF SATURATION, the upper limit of which is the WATER TABLE. (⟡UNSATURATED ZONE.)

Agate. ⟡SILICA.

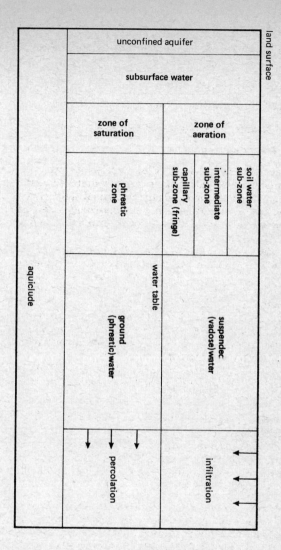

Figure 7. Idealized vertical section through the zones of aeration and saturation in an unconfined aquifer with the process of subsurface water movement.

Agglomerate. ⟡PYROCLASTIC ROCKS.

Aggregate. An aggregate in the earth resources' sense is a bulk material, generally of low monetary or unit value, which is used in the construction industry. Heavy aggregates include gravel, sand, crushed stone and some crushed brick. Lightweight aggregates include cinders, sintered fuel ash, coke breeze, clinker, foamed slag, pumice, diatomite, coral and EXPANDABLE ROCKS AND MINERALS. Heavy aggregates are employed for strength and durability, while lightweight aggregates are used where lower-density materials are required, especially when heat and sound insulation properties are important.

Fine aggregate will pass through a $\frac{3}{16}$-inch mesh, and is mainly sand. Sand containing angular quartz grains, called sharp sand, is particularly suitable for making CONCRETE. Soft sand contains rounded grains; being more workable than sharp sand, it is mainly used for making mortar.

Coarse aggregate will not pass through a $\frac{3}{16}$-inch mesh. Gravel is a natural coarse aggregate, though it commonly contains sharp sand in addition to pebbles. Much coarse aggregate is CRUSHED STONE.

Concrete which is to be used for a specialized purpose, such as nuclear shielding, is increased in density by the addition of higher-density materials. The addition of barite or limonite results in a relatively small density increase, magnetite or ilmenite produce a moderate density increase, and for a high density increase ferrophosphorus or steel are added.

Aggregate for making concrete should be free, or nearly free, of organic material, dust, feldspar, silt, clay, and iron pyrite. It should be well graded and composed of durable, non-porous, non-swelling rocks. Crushed stone which is comprised of angular rock fragments forms a better bond with bitumen than do rounded pebbles from gravels. Thus, for road-making, crushed stone is generally preferred. Poor-quality aggregates unsuitable for making concrete may be of value as hardcore or fill.

In most countries the demand for aggregates is continually rising. This leads to conflicts in land use. For example, in Britain, where sand and gravel deposits are mainly restricted to lowlying land in the south and east, their exploitation for aggregate conflicts with the use of the same land for agriculture, housing, industrial development and water supply. Old gravel pits may possess an amenity value as sites for water sports or nature reserves. Many of the hard rocks suitable for the preparation of crushed stone crop out in upland regions where their exploitation conflicts with landscape value. The demand for aggregate is such that submarine deposits are being actively explored.

Air drilling. A method of rotary drilling using compressed air to transport the rock cuttings from depth in the well to the surface. It is a relatively modern development within the water-well industry. Air is circulated at high velocity through the drill pipe, out through ports in the drill BIT, and upward in the annular space around the drill pipe to carry the rock cuttings to the surface or blow them into rock crevices. ($\Diamond$DRILLING METHODS.)

Alabaster. $\Diamond$GYPSUM.

Albite. $\Diamond$FELDSPAR.

Alkali lake. Alkali lakes develop in regions of hot, arid climate where lakes are fed by streams with no outlet from the lake. They differ from SALINE LAKES in that they are richer in the dissolved carbonates of sodium or potassium than in chlorides, although they may contain some chlorides, sulphates, or even borates. ($\Diamond$BRINE.)

Alkalinity. Alkalinity in water is its ability to neutralize acid and is usually expressed as calcium carbonate equivalents. It does not necessarily mean that the water has a pH value ($\Diamond$pH) above that of 'neutral' water which is 7. Ground water with a pH value below 7 may, at the same time, possess some salts that will neutralize acid and thus will have some measurable alkalinity. Carbonate and bicarbonate ions in water contribute to alkalinity. Chloride, sulphate and nitrate ions do not. ($\Diamond$ACIDITY.)

Allophane. $\Diamond$CLAY MINERALS.

Alluvial deposit. Detrital rock material which has been transported and deposited by rivers and streams, commonly composed of gravel, sand or silt. It may contain local concentrations of some valuable mineral, e.g. gold, diamonds, platinum or tin. ($\Diamond$PLACER DEPOSIT.) (Figure 8.)

Alum. The highly soluble minerals of the alum group are hydrated alkali alumino-sulphates, the one usually known as alum being potash alum ($KAl(SO)_2.12H_2O$). The mineral alunite, alum rock or alumstone ($K_2Al_6(OH)_{12}(SO_4)_4$) has for long been used for the manufacture of alum. Alum generally occurs as an efflorescence in clay rocks. It results from the reaction of sulphuric acid, produced during the decomposition of iron pyrites, with alkali-rich aluminio-silicates, such as some of the clay minerals. Alunite is usually associated with volcanic rocks which have been altered by sulphurous gases. An ammonia-rich alum is associated with some brown coals and bituminous shales. A soda alum also occurs in some shales. Alums are used in water purification, sizing of paper, in medicine, in baking powder, in dyeing, and in many minor miscellaneous industrial processes. Alunite is a potential source of POTASH. Some alum is made from bauxite.

Aluminium (Al). The third most abundant element (8·13 per cent) in the earth's crust. Next to iron, it is the most widely used metal. It is a good conductor of electricity, is light in weight and forms high-strength alloys in combination with other metals. Aluminium is widely used in the construction industry, the manufacture of motor cars and aeroplanes, electrical transmission, packaging (aluminium foil) and home consumer products.

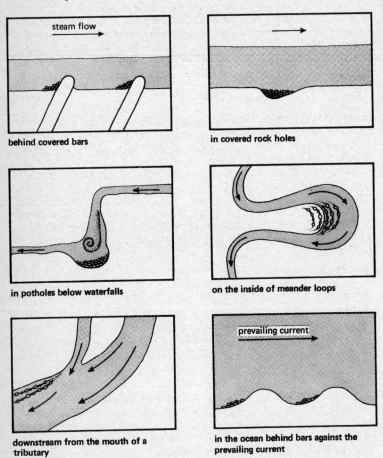

behind covered bars

in covered rock holes

in potholes below waterfalls

on the inside of meander loops

downstream from the mouth of a tributary

in the ocean behind bars against the prevailing current

Figure 8. Alluvial or placer deposits. These occur where an obstacle or change of gradient occurs in the bed of flowing water, as in rivers or at the mouths of streams and, under some circumstances, on the sea floor. (After Skinner.)

A variety of minerals is used as a source of aluminium and there is no shortage of potential ore minerals; eventually the metal may be produced from ordinary clay. At the present time, however, almost all commercial aluminium is derived from bauxite. The process involves treatment of the ore (Bayer process) to produce alumina and then electrolytic reduction ('smelting') of the oxide. Depending on the grade of ore, between 4 and 6 tonnes of bauxite are required to produce 1 tonne of aluminium. Production of primary aluminium demands very large amounts of electrical power. The principal producers of primary aluminium in 1974 were the USA, USSR, Japan and Canada. Total world production in 1974 was 13·7 million tonnes.

Alunite. ⟡ALUM.

Amosite. ⟡AMPHIBOLE; ASBESTOS.

Amphibole. Most of the minerals belonging to the amphibole group are hydrous ferromagnesian silicates. They are mainly dark-coloured, and occur in intermediate igneous rocks and some metamorphic rocks. The best-known amphibole is hornblende, a mineral found in some syenites and in amphibolite, a metamorphic rock containing hornblende plus plagioclase feldspar. Among the amphiboles of commercial importance are actinolite, amosite, anthophyllite, crocidolite and tremolite, all of which occur in some ASBESTOS deposits. Amphiboles are also associated with some TALC deposits, and some of them are potential sources of raw material for the manufacture of MINERAL WOOL.

Amphibolite. ⟡AMPHIBOLE.

Andalusite. ⟡SILLIMANITE GROUP OF MINERALS.

Andesite. A fine-grained volcanic or DYKE rock of intermediate composition containing the minerals plagioclase FELDSPAR and hornblende (⟡AMPHIBOLE) or augite (⟡PYROXENE). It is equivalent in composition to DIORITE, a coarse-grained igneous rock. Because andesites are generally closely jointed they are of little use except for CRUSHED STONE.

Anglesite. A soft (hardness 2·5–3), white or pale-coloured lead ore mineral ($PbSO_4$) with a high (6·2–6·4) SPECIFIC GRAVITY. When crystalline it has an adamantine lustre but it frequently occurs in earth-like masses. Anglesite is found in the zone of oxidation above veins of galena. It contains 68 per cent lead.

Anhydrite. A white mineral ($CaSO_4$) which generally occurs in association with GYPSUM and other minerals characteristic of EVAPORITE DEPOSITS formed when a body of seawater evaporates. Anhydrite, rather than gypsum, is crystallized when the temperature of the saline

27

water exceeds about 25°C, when salt crystallization has started, or when gypsum crystallization is complete. Some primary gypsum is altered to anhydrite by dehydration following burial. Anhydrite also develops in the CAP ROCK over salt intrusions.

In Britain anhydrite is worked from rocks of Permian age in the north of England.

It is used in the manufacture of sulphuric acid and some nitrogenous FERTILIZERS; it also has a limited use in the making of plasters and CEMENTS.

Anorthite. ◊FELDSPAR.

Anorthosite. A rare, coarse-grained, plutonic igneous rock which is comprised almost entirely of calcic-plagioclase FELDSPAR. It can be thought of as a GABBRO which is free of mafic (◊IGNEOUS ROCKS) minerals such as augite. There has been some investigation of anorthosites as sources of feldspar, especially for use in paints employed for road-marking.

Anthophyllite. ◊AMPHIBOLE; ASBESTOS.

Anthracite. ◊COAL.

Anticline. ◊FOLD.

Antimonite. ◊ANTIMONY.

Antimony (Sb). A soft, white and brittle metal which is used as a constituent of metallic (especially lead) alloys and incorporated in the manufacture of storage batteries, ammunition and telecommunication equipment. Antimony compounds have a variety of uses including flame-proofing of textiles, as pigments and in the production of glass and ceramics.

A nation's consumption or stockpiling of antimony dramatically increases in times of war or if it anticipates war. (◊STRATEGIC MINERAL.)

Antimony (crustal abundance 0·2 ppm) occurs native and forms a large number of compounds, but only stibnite or antimonite (Sb_2S_3) is important as an ore. It occurs in quartz veins associated with granitic rocks. A substantial, but difficult to document, amount of commercial antimony is recovered as a by-product during the processing of other BASE METAL and gold–tungsten ores. Recycling of scrap antimony (especially battery plates) is extensively carried out. (◊◊BACTERIAL EXTRACTION OF METALS.)

Accurate statistics for antimony are difficult to obtain as the Communist bloc, especially China and Russia, controls nearly a third of the world's production. In 1974 it was of the order of 71000 tonnes. Over 20 per cent of this came from the Consolidated Murchison Mine in the East Transvaal, South Africa.

Apatite. ⋗PHOSPHATE DEPOSITS.

Aquiclude. Relatively impermeable material which may contain or store water but is incapable of transmitting water in significant quantities. Clay, for example, may possess POROSITY values in excess of 40 per cent but the transmissive nature of this fine-grained material, expressed as permeability, may be as low as 10^{-3} to 10^{-5} darcys (⋗DARCY'S LAW). In the subsurface, aquifers (permeable formations) are normally confined by underlying and overlying aquicludes (Figure 9). (⋗AQUITARD.)

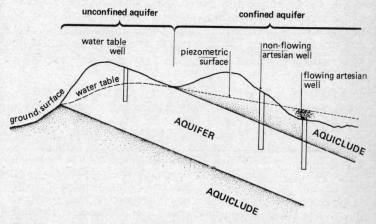

Figure 9.   Unconfined and confined aquifer conditions.

Aquifer. Any material that yields water readily enough to be significant as a source of supply. An aquifer serves as both a subsurface reservoir and a conduit, since it has the ability to store and transmit water.

Aquifers can comprise any unconsolidated or consolidated sedimentary sequences such as sands, gravels, sandstones, conglomerates and limestones. Aquifers can also include fractured zones in dense plutonic igneous rocks or openings in volcanic rocks as in lavas. Rarely are aquifers homogeneous with regard to their overall rock properties, and they can occur in an almost infinite variety of shapes with varying water-bearing properties.

Ground water may leak out of aquifers into adjacent AQUICLUDES but may also receive water from aquicludes. The passage of poor-quality water from the surface via fissures or fractures direct to the

ZONE OF SATURATION of an aquifer can lead to pollution of the contained ground waters.

Subsurface, CONFINED (ARTESIAN) AQUIFERS may be compressed by the superincumbent load following withdrawal of ground water from such an aquifer. Compaction of aquifers undergoing non-elastic compression may in the long term give rise to serious land subsidence problems. Many areas of surface land subsidence due to deep ground water pumping are known in the USA – particularly California – Japan, Mexico and England. Possibly the most publicized example of this phenomenon in recent years has been the subsidence problems of Venice.

(◊COASTAL AQUIFER; LIMESTONE AQUIFERS; PERCHED AQUIFER; UNCONFINED AQUIFER.)

Aquifuge. A totally impermeable rock which neither contains nor transmits water. Unjointed, solid granite and unfissured, quartz-cemented sandstones and cherts are good examples of such material. (◊AQUITARD.)

Aquitard. Naturally occurring material (of a semi-confined nature), which can store water but only permits some ground water flow at a very slow rate. Thus movement of ground water through aquitards may be significant in regional migration of water but locally would not yield significant supplies of water to individual wells.

AQUIFER, AQUICLUDE, AQUIFUGE and aquitard thus all lack precise definitions with respect to measurable physical properties. For example, an aquifer that may produce an individual well yield of 100 gallons per day in a desert region would be classed as an aquitard, or even an aquiclude, in a more temperate climatic area where individual wells in a valley filled with gravel may produce several million gallons per day.

Aragonite. ◊CALCITE.

Arenaceous rocks. ◊SEDIMENTARY ROCKS.

Argentite. ◊SILVER.

Argillaceous rocks. ◊SEDIMENTARY ROCKS.

Arkose. ◊SANDSTONE.

Arsenic (As). The chief uses of arsenic are as a constituent in agricultural pesticides, herbicides or defoliants and as a decolorizer in GLASS-making. Arsenic is commercially recovered only as a by-product during the smelting of other, especially BASE METAL, ores. Sweden is the major producer.

Artesian well. When a CONFINED (ARTESIAN) AQUIFER is pierced by a well the ground water will rise to a height in that well above the top of the aquifer. If the PIEZOMETRIC SURFACE is below ground level, then the level to which the water will rise will also be below ground surface. Such a well would be a non-flowing artesian well (Figure 9). Conversely, if the piezometric surface is above ground level the water will overflow from the well at the ground surface. Such a condition is described as a flowing or overflowing artesian well (Figure 9).

The term 'artesian' is derived from the northern French province of Artois where the first deep wells drilled to tap confined aquifers date from about 1750.

Artificial recharge. Artificial ground water recharge can modify the HYDROLOGICAL CYCLE by introducing ground water to an aquifer in excess of natural replenishment. This augmentation above the natural INFILTRATION or PRECIPITATION of surface water into aquifers is undertaken by some method of construction, spreading of water (IRRIGATION), or by artificially changing natural conditions. Thus the artificial replenishment of an aquifer may involve one of a variety of methods, including WATER SPREADING, recharging through pits, excavations, wells and shafts, and pumping to induce recharge from surface water bodies.

Artificial ground water recharge is not as yet extensive, but as the demand for water in heavily populated countries increases this practice may be an important key in water resources development. West Germany obtains more than 30 per cent of its public water supply by artificial ground water recharge practice. (◇ RECHARGE AREA.)

Artificial stone. Artificial or reconstituted stone, a type of CONCRETE, is made by recementing granulated rock. Compared with DIMENSION STONE its advantages are relative cheapness, the ability to cast it into moulds of predetermined dimensions, and that a continuous supply of material of the same appearance can be guaranteed. Rocks which are too closely fractured for use as dimension stone can be worked. Artificial stone can be cast into reinforced beams.

Asbestos. A group of fibrous minerals valued because of their great resistance to fire, heat, friction and chemical decomposition, their low electrical conductivity, sound insulation properties, high tensile strength, and their ability to be spun.

In terms of tonnage the most important variety of asbestos is serpentine asbestos, the mineral chrysotile ($Mg_6Si_4O_{10}(OH)_8$). Its fibres are flexible, silky, capable of withstanding very high temperatures and of high tensile strength. Chrysotile deposits occur in association with the altered ultrabasic igneous rock known as SERPENTINE. Some chrysotile is also associated with altered impure dolomitic limestones which have been metamorphosed and contain appreciable quantities of the mineral OLIVINE. The best-known deposit of chrysotile occurs in the Thetford District of Quebec Province, Canada. Chrysotile is also worked in the Urals of Russia, and in Rhodesia. Despite its many valuable properties, chrysotile is not an acid-resistant variety of asbestos.

The AMPHIBOLE asbestos group of minerals includes: anthophyllite $(Mg_7Si_8O_{22}(OH)_2)$, which occurs in some altered metamorphic lime-stones, is resistant to acids, is brittle, and of low tensile strength; amosite, an iron-rich variety of anthophyllite, which yields unusually long fibres but which is attacked by acids; crocidolite, a fibrous variety of the mineral riebeckite $(Na_2Fe_3{}^{+2}Fe_2{}^{+3}(Si_8O_{22})(OH)_2)$, which is less resistant to fusion but which is resistant to acids; fibrous tremolite $(Ca_2Mg_5(Si_8O_{22})(OH)_2)$, which is brittle and of low tensile strength but which like tremolite occurs in association with TALC in altered impure metamorphosed limestones. Amosite and crocidolite are mined in South Africa. Actinolite is rarely used alone but may be mixed with other asbestos minerals.

Considering the properties of many of the asbestos minerals, it is not surprising that many are used for manufacturing fireproof materials, brake linings and clutch facings. Some, especially anthophyllite, are used for the electrical insulation of cables and switchboards; crocidolite is used for filtering acids. Asbestos dust is considered to be a serious health hazard. Production in 1972, in tonnes, was: Canada, 1535000; USSR, 1220000; Brazil, 360000. (⟡MINERAL FILLERS.)

Ash.   ⟡PYROCLASTIC ROCKS.

Asphalt.   Highly viscous or solid natural PETROLEUM compounds, known as asphalt, bitumen, pitch or tar, tend to flow at high tempera-tures, and to remain as a residual material after other petroleum sub-stances have evaporated. Asphalt was used as a mortar between bricks by the Sumerians (in what is now Iraq) about 5500 BC. It is now primarily in demand as a material for road construction but is also used in the manufacture of roofing materials (⟡GRANULES).

Asphalt lakes are surface accumulations of solid and semi-solid asphaltic substances representing the final stages in the natural destruc-tion of an original crude oil reservoir. Oxidation and evaporation remove the more volatile constituents of the crude oil and the mass may be replenished gradually from below. The most famous is the Trinidad Pitch Lake but others are known in California, the Middle East and the USSR.

Assay.   Used to determine the amount of sought material (usually metal) in an ore. An assay differs from a complete chemical analysis in that, usually, only certain elements are determined.

An assay ton is a sample of 32·667 grammes for a long ton of 2240 pounds avoirdupois or 29·1667 grammes for a short ton (2000 pounds avoirdupois). The number of milligrammes of precious metal in an assay ton equals the number of troy ounces of precious metal per ton of ore.

Asthenosphere. ⟡THE EARTH.

Atmosphere. The envelope of air (and water vapour) which surrounds the earth, consisting of nitrogen 78+ per cent, oxygen 20·9 per cent, argon 0·93 per cent and carbon dioxide 0·03 per cent by volume, plus traces of neon, helium, krypton, xenon and ozone. Water vapour, ozone and carbon dioxide absorb infra-red radiation emitted from the ground and affect atmospheric temperatures. The atmosphere is a commercial source of gases which are obtained by liquefying air and controlling the temperature so that only the required gas is liberated.

Attapulgite. ⟡ADSORBENT AND BLEACHING CLAYS; CLAY MIN-ERALS.

Augite. ⟡PYROXENE.

Axial plane. ⟡FOLD.

Azurite. A hydrated copper carbonate mineral ($Cu(CO_3)_2(OH)_2$) containing 55 per cent copper. It is found in association with other copper minerals in the zone of oxidation, and is characterized by its beautiful azure blue colour

B

Bacterial extraction of metals. Certain bacteria are capable of oxidizing iron and some other ores so that a rich solution of the metal is produced. These processes take place very slowly in nature, but experiments in artificially speeding them up to make a commercially exploitable method of ore treatment by bacteria are in progress. The results are encouraging, and ores of uranium, nickel, copper and antimony in addition to iron can all be treated in this way. Bacterial treatment of mine wastes may also lead to the recovery of metallic oxides not considered worth processing by other means at present.

Ballas. ♢DIAMOND.

Ball clay. Clays composed principally of the CLAY MINERAL kaolin which unlike the kaolin in CHINA CLAY is less well crystallized. They are especially 'fat' (plastic), and particularly suitable for the manufacture of PORCELAIN and WHITEWARE.

British ball clays of Tertiary age are worked in the Bovey Tracey Basin of Devon, and in Dorset. Cretaceous ball clays are exploited in New Jersey, and Tertiary ball clays in Kentucky and Tennessee, USA.

Banded iron formations. Ancient sedimentary deposits consisting of a repeated alternation of thin iron-rich and silica-rich (chert) layers. They are invariably Precambrian in age (clustering about 2200 million years) and are found on all continents.

The banded iron formations (synonyms include taconite and itabarite) represent an almost inexhaustible supply of IRON ore. Any of the major iron minerals may occur but HEMATITE and MAGNETITE are commonest. The ores are low grade (iron content 15–40 per cent) and were originally mined only where SECONDARY ENRICHMENT has occurred. However, modern methods of ORE DRESSING, especially agglomerating the finely ground ore into pellets, can now upgrade the ore to an iron content of more than 60 per cent. Such pelletized ore is often preferred to traditional ores.

Bank storage. During a FLOOD period of a surface stream when the RIVER stage is high, ground water levels in adjacent permeable deposits are raised temporarily by inflow of water from the stream. The area next to the river channel will experience the maximum rise in ground water levels. This water is thus temporarily stored in the permeable deposits (and when the river stage falls the water drains back to the river). This process is known as bank storage and the rate of inflow and outflow may be measured. (Figure 25.)

Barite. ◊BARIUM MINERALS.

Barium minerals. The two minerals from which barium (Ba) compounds are obtained are barite ($BaSO_4$) and witherite ($BaCO_3$). Barite, also known as barytes, is more abundant than witherite. It is called heavy-spar by some miners because of its relatively high density (4·3–4·6 grammes per cubic centimetre). When pure, barite is colourless, but it is often white or coloured yellow or brown by impurities. Most commercial sources of barite are veins of hydrothermal origin where the barite may occur either as the principal economic constituent or in association with metallic sulphides, as in the North Pennine orefield in Britain. Barite is also associated with a nepheline SYENITE in Arkansas. Minerals commonly found with barite include CALCITE, quartz (◊SILICA), FLUORITE, GALENA and sphalerite (◊ZINC). Barite also occurs in some clays where it appears to have grown in place, and some is found replacing calcite in limestones. Deposits of barite in Missouri which are of residual origin were probably derived from the weathering of limestones. In Britain the only significant commercial source of witherite occurs in Northumberland. Witherite may result from the secondary alteration of barite.

Barite is used as an extender in the manufacture of paint and heavy paper, and as a MINERAL FILLER in the making of rubber, linoleum and oilcloth. Crushed, low-grade barite is added to DRILLING MUD to increase its density, and thus prevent blow-outs in oil-well drilling. Barium hydrate is employed in sugar refining, and barium nitrate is used for producing green colours in fireworks. Barite may be added to CONCRETE when its density has to be increased and when it is to be used to screen radiation. Barite is also used in GLASS-making as a flux and to improve the brilliance of the glass. The employment of 'barium meal' in medicine is well known. Although less abundant than barite, witherite is a suitable alternative raw material. Barite production, in 1974, in short tons was: USA, 1104000; West Germany, 400000; Ireland, 325000.

Barrel. The standard volumetric unit of crude oil, equivalent to 42 US gallons or 34·97 Imperial gallons, approximately 310 lb. It produces between 5400000 and 6000000 Btu (British thermal units) of energy.

Barytes. ◊BARIUM MINERALS.

Basalt. A dark-coloured, fine-grained basic volcanic rock containing calcic-plagioclase FELDSPAR, augite or other PYROXENES, and in some basalts OLIVINE. It is the fine-grained equivalent of GABBRO, a coarse-grained igneous rock. Basalt is found in many plateau lava sequences which have erupted from fissure vents. Some basalts form DYKES or SILLS. Many basalts are closely jointed; commonly the joints are arranged so as to give columns, as for example at the Giant's Causeway in Northern Ireland. The close fracturing of most basalts makes them suitable for use as CRUSHED STONE. Basalt and DOLERITE are particularly valuable for road making because they adhere well with asphalt. (◊ROAD METAL). It is also used as an ABRASIVE. (◊GRANULES.)

Base exchange. Very fine-grained material such as silt, clay or organic particles in an aquifer may adsorb and retain cations on their surfaces. The cations, which are held by minute electrical charges at the surface of the particles, can be replaced by others present in the ground water as it flows through the aquifer. This reaction is known as base, or cationic, exchange and is a most important factor in controlling the chemistry of ground waters for public water supply at individual well sites.

Sodium, magnesium and calcium are the principal cations involved, and ion-exchange reactions are reversible. In LIMESTONE AQUIFERS the ground waters directly underlying the RECHARGE (outcrop) AREA of the aquifer are dominated by calcium and magnesium cations and carbonate, bicarbonate anions. This results in a 'hard' water (◊HARD-NESS (1)). As the ground water flows downdip through the aquifer, where covered by a clay AQUICLUDE, base exchange takes place with sodium replacing the calcium and magnesium cations and producing a 'soft' water zone. Such zones can be recognized in the chalk aquifer of the London Basin and the limestone aquifer of Lincolnshire.

Base flow. That portion of river flow contributed by ground water discharge. It therefore represents the ground water component comprising sustained fair-weather RUNOFF. In certain rivers base flow may be a negligible amount of the total flow under high discharge conditions and, conversely, may represent the total river flow during drought periods. The base flow contribution to river discharge is commonly steady throughout the year and contrasts with the rapid fluctuations seen in the surface runoff component to river discharge over the same time (Figure 31).

The amount of base flow in a stream may be found from the river discharge HYDROGRAPH. In permeable catchments, where development of an aquifer is being carried out via pumping from wells, any base flow in the river may be regarded as surplus ground water resources.

Base metal. To the alchemist, literally any non-precious metal. In the mining industry, it refers especially to the ores of copper (Cu), lead

(Pb), zinc (Zn), tin (Sn), mercury (Hg) and antimony (Sb). (⟡FERROUS MINERALS; PRECIOUS METALS.)

Basic rocks. ⟡IGNEOUS ROCKS.

Bastunaesite. ⟡RARE EARTH ELEMENTS.

Batholith. An igneous intrusion of at least 100 square kilometres in extent (also batholite, bathylith). Some batholiths are hundreds or thousands of kilometres in length. The largest are generally composite, comprising several smaller bodies intruded into each other. Although batholiths cut across the rocks into which they are intruded, many major batholiths are elongate parallel to the fold mountain belts in which they occur. Uplift and subsequent denudation are responsible for the rocks of batholiths being exposed at the present-day surface (Figure 10). Where they are incompletely unroofed, only the higher parts, or cupolas, are exposed. Stocks and bosses are smaller intrusions.

The plutonic igneous rocks of batholiths are generally coarse-grained on account of relatively slow cooling. The commonest rock in large batholiths is granite, but some smaller batholiths contain diorite or gabbro.

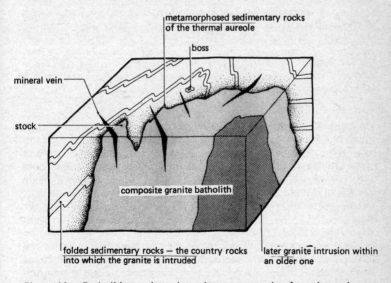

Figure 10. Batholiths are large intrusions, commonly of granite rock, into deeply buried country rocks. Their formation is accompanied by great heat and pressure. Many small intrusions and mineral bodies may be associated with them.

The Dartmoor granite is the extreme eastern cupola of a major batholith which underlies, at a depth not generally exceeding 1 kilometre, south-west England from the Scilly Isles to Dartmoor. Other cupolas that rise above the general level of this largely unexposed batholith are those of Land's End, Bodmin and St Austell.

Batholiths are of interest to some economic geologists because many metallic, and some non-metallic, mineral deposits are associated with them.

Bauxite. The principal source of ALUMINIUM, bauxite is a pale-coloured earthy mixture of several hydrated aluminous minerals ($Al_2O_3n.H_2O$) associated with minor amounts of other minerals (diaspore).

It is a residual deposit formed, in recent or past times, under tropical or sub-tropical conditions, by prolonged weathering and leaching of aluminium-bearing rocks. Bauxite is widely distributed but the major deposits are concentrated in the less industrialized parts of the world away from the main consuming and smelting centres in North America, Europe and Japan. The Intergovernmental Association of Bauxite Producing Countries (Australia, Jamaica, Surinam, Guyana, Guinea, Sierra Leone and Yugoslavia) produce more than three-quarters of the free-world production. The USSR is also a major producer. Total world production of bauxite in 1974 was 73 million tonnes. (⟡LATERITE.)

Bed (of rock). The smallest commonly recognized unit of stratified (or bedded) rock, e.g. a coal seam or any rock unit that is separated from others above or below by a parting or bedding plane.

Bentonite. A clay which contains a high proportion of the CLAY MINERAL montmorillonite. Bentonites containing montmorillonite in which sodium is the principal exchangeable ion are known as 'Wyoming', 'Western' or 'True' bentonites. Because these clays swell in water and form thixotropic gels (⟡THIXOTROPY) they are also known as swelling-type bentonites. The 'Southern' or sub-bentonites contain montmorillonite in which calcium is the principal exchangeable ion. These are the non-swelling bentonites. There are gradations between the two types.

It is thought that bentonites were originally deposited as airborne volcanic ash, and were converted to clay by devitrification and the chemical action of saline and ground waters. Other bentonites, such as those in Mississippi, were probably developed elsewhere and transported as detritus to their present sites. Many are of Mesozoic or Tertiary age.

Although the usage of the name Fuller's Earth is not always consistent, it is commonly employed to describe naturally active, non-swelling, magnesium-rich montmorillonitic clays. In the south of

England there are commercially important deposits of Fuller's Earth in Middle Jurassic and Lower Cretaceous rocks.

Non-swelling bentonites are used as ADSORBENT AND BLEACHING CLAYS and as CATALYTIC CLAYS. Swelling bentonites are used as DRILLING MUD CLAYS, as BONDING CLAYS, and for sealing dams, reservoirs and ditches. Bentonites are also used in the manufacture or processing of fungicides and insecticides, as mild ABRASIVES, and in the chemical and cosmetic industries.

The principal area in the USA where sodium bentonites are worked is Wyoming and adjacent states. Calcium bentonites and intermediate types of non-swelling bentonites are exploited in Mississippi, Texas and Arizona. Elsewhere bentonites are worked in Germany, Hungary, Mexico, South America, Italy, New Zealand and England.

Beryllium and beryl. Beryllium alumino-silicate, the mineral beryl ($3BeO.Al_2O_3.6SiO_2$), is concentrated in commercial quantities in some granite PEGMATITES and veins. It occurs as prismatic, hexagonal crystals up to about 1 metre in length, and is generally a pale translucent blue or green colour. Beryl is commonly obtained as a by-product of working pegmatites exploited primarily for mica or feldspar. Beryl is the principal source of the element beryllium (Be) which is employed as a hardening agent in some alloys. Beryllium oxide is used as a high-temperature REFRACTORY material. Some beryllium compounds are highly toxic. Total world production (1968) was in the order of 200 tonnes in metal content.

Biotite. ⊳MICA.

Bismuth (Bi). A rare element (crustal abundance 0·2 ppm), but native bismuth and various bismuth compounds are common trace minerals in many base metal ores. Commercially, bismuth is produced as a by-product of the processing of lead and copper ores. It is used principally in pharmaceuticals, fusible and other alloys. Only unreliable statistics are available; Peru, Bolivia, Mexico and Japan are major producers. Total Western world production (1974) is estimated at 9 million lb.

Bit. The drilling tool that actually cuts the hole in the rock. Most bits used in drilling for oil and minerals have diamonds embedded in the cutting edges. (Figure 11.)

Bittern. A bitter-tasting liquid which remains after salt has been crystallized from a highly saline liquid or brine.

Bitumen. ⊳ASPHALT.

Bituminous coal. ⊳COAL.

Bleaching clays (bleaching earth). ⊳ADSORBENT AND BLEACHING CLAYS.

Bloatable clays. ⊳EXPANDABLE ROCKS AND MINERALS.

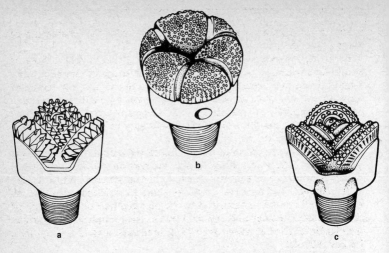

Figure 11. The bit is the essential drilling tool in the sinking of boreholes. The cutting edges may be of hard metals or of diamonds. Bits range in size from a few centimetres to nearly 1 metre in diameter. Three types are shown: (a) for soft rock formations; (b) a diamond-edged bit; (c) a tungsten-carbide 'button' bit for hard formations.

Blue John. ⬦FLUORITE.

Bonding clays. Clays used in foundries to bind sand together into the desired shape of a mould. The principal types used are BENTONITES and FIRECLAYS. Two or three times the quantity of fireclay may be required compared with bentonite. In the past some natural deposits which are a mixture of sand and clay were widely used, these being known as naturally bonded sands.

Borate deposits. Boron (B) and compounds derived from it are extracted from four evaporite minerals (⬦EVAPORITE DEPOSITS): borax ($Na_2B_4O_7.10H_2O$), kernite ($Na_2O.2B_2O_3.4H_2O$), colemanite ($Ca_2B_6O_{11}.5H_2O$) and ulexite ($NaCaB_5O_9.8H_2O$). Minor quantities of borates are obtained from natural boric acid or sassolite ($B(OH)_3$), a relatively rare mineral.

The majority of borate deposits occur in rocks of Cenozoic age and of continental origin in volcanic areas where the climate is particularly arid.

The principal producer of boron minerals in the Western world is the USA. The important borax deposits of Kern County in the Mojave Desert of California occur at depths greater than 100 metres in rocks which were laid down in a temporary lake during the Pliocene. The boron in this deposit was probably derived from the weathering of nearby lavas or from volcanic gases.

Borate deposits outside the USA are more varied in type – those of South America, for example, are near-surface deposits in which other evaporite minerals occur. Sassolite is worked in the Tuscany region of Italy from steam vents associated with past volcanic activity. Permian potash deposits at Stassfurt in Germany contain boracite ($Mg_6Cl_2B_{14}O_{26}$), resulting from the final stages of the drying up of a body of seawater.

The principal use of borax is as a flux in the manufacture of GLASS and enamel, when according to the proportion of borax added it improves brilliance or heat resistance. Other uses are as an insecticide, water softener, and preservative of fruit and timber. It is also used as an additive in FERTILIZERS and in the manufacture or processing of paper, textiles, leather and paint. Borax is also employed in various chemical industries. Production in 1971, in tonnes, was: USA, 950000; Turkey, 400000.

Borax. ◊BORATE DEPOSITS; SODIUM CARBONATE AND SULPHATE MINERALS.

Bornite. An important copper ore. It is a sulphide of copper and iron (Cu_5FeS_4), containing 63 per cent copper. When fresh, the mineral is bronze in colour but it tarnishes on exposure to air and has a characteristic, often iridescent, purplish colour. Bornite is usually found in association with CHALCOCITE and CHALCOPYRITE.

Boron. ◊BORATE DEPOSITS.

Bort. ◊DIAMOND.

Brackish water. Contains between 1000 and 10000 ppm of total dissolved solids. It contrasts with potable freshwaters which contain between 0 and 1000 ppm of total dissolved solids. Brackish or moderately salty ground waters exist in extensive aquifers in many places. Although unfit for drinking purposes at its present level of total dissolved solids, brackish water may be used in industry. DESALINATION methods may in future become cheap enough to make use of originally brackish water in public water supplies.

Breccia. ◊CONGLOMERATE.

Brick clays. CLAYS or mudstones (◊MUD AND MUDSTONE) used for making bricks which are commonly of low grade and similar to those employed in the manufacture of other STRUCTURAL CLAY PRODUCTS.

The three categories of bricks are: concrete bricks; sand-lime and calc-silicate bricks; and common fired-clay bricks. The two properties of clay which make it suitable for brick-making are its plasticity when moist, and its ability to harden into material of high strength when heated into its VITRIFICATION range.

Although the specifications that a clay or mudstone should fulfil when it is to be used for common brick-making are not rigorous, deposits of suitable clay should be uniform, largely free of excess organic material, gravel-free (or capable of being cheaply separated from the gravel), largely non-calcareous, and free of iron pyrites.

Impurities in the clay may give the bricks made from them a distinctive colour, for example, ferric oxides impart a red colour. Refractory bricks are made from REFRACTORY CLAYS. High-quality engineering bricks are commonly made from higher-grade clays or mudstones, especially those rich in silica or with admixtures of siltstone.

In many hot arid areas of the world sun-dried bricks (adobe) were (and sometimes still are) widely used. The clay used for making these bricks is often mixed with sand and/or straw.

The Thanet sand (of Eocene age) and the Chalk (of Cretaceous age) have been mixed to make sand-lime bricks. The name brick-earth is used for a loam (of Pleistocene age) employed for making bricks in the Thames Valley.

Clays of suitable quality for brick-making are common in most countries of the world. In Britain clays or mudstones for brick-making have been worked from the rocks of most geological periods younger than the Silurian. Some well-known deposits worked for brick clay include the coal measures, the red Keuper mudstones, the Liassic, Oxford and Kimmeridge clays, the Weald and Gault clays, the Reading Beds and the London Clay, and several Quaternary clays including boulder clay.

Brine. Water containing more than 100 000 ppm of total dissolved solids from which salts may be obtained after evaporation. Some brines are the result of ground water flow through EVAPORITE DEPOSITS, while others are CONNATE WATER. Subsurface deposits are exploited by pumping water through them to yield artificial brines. Some ALKALI and SALINE LAKES yield brines of value. Brines are associated with many oilfields.

Brines may yield salts or elements discussed under the headings of BORATE DEPOSITS; BROMINE SOURCES; IODINE SOURCES; MAGNESIUM SALTS; NITRATE DEPOSITS; POTASH DEPOSITS; SALT; and SODIUM CARBONATE AND SULPHATE MINERALS.

Bromine sources. Bromine (Br), which is a corrosive liquid at room temperature, is generally extracted as bromides from seawater BRINES, and from some SALINE LAKES. The Dead Sea in Palestine has been one source. Bromides have also been obtained as a by-product during the working of the POTASH DEPOSITS of Stassfurt in Germany.

Bromine compounds are used in medicine and photography, but most important in terms of volume is ethylene dibromide which is used as an additive in the manufacture of anti-knock petrol (gasoline). Bromine compounds are used for fumigation and sterilization, and in many manufacturing processes.

Building stone. ⇨DIMENSION STONE.

Burtonization. ⇨GYPSUM.

By-product. A metal or mineral produced incidentally to the main sought-after product. The by-product may be produced during the mining, or during the subsequent processing and refining of the ore.

C

Cable-tool drilling. ◊DRILLING METHODS.

Cadmium (Cd).   A relatively rare element with a crustal abundance of 0·2 ppm. Its principal use is in the electroplating of metals to protect them from corrosion; it is also used in the production of pigments (cadmium yellow), batteries, electronic instruments and plastic materials.

Cadmium is found in nature as the sulphide mineral, greenockite, but this is of rare occurrence. All the cadmium of commerce is recovered as an industrial by-product during the smelting and refining of ZINC and other BASE METAL ores. World production in 1974 was estimated at 23·8 million lb.

Calcarenites. ◊SANDSTONE.

Calcite.   A mineral, calcium carbonate ($CaCO_3$), crystallizing in the trigonal system, occurs as colourless or white crystals of hardness 3 on Mohs' scale. It is the principal mineral in LIMESTONES; it also occurs in marls (◊MARL AND MARLSTONE), and as a natural cement in many SANDSTONES. Calcite is also a common mineral in many veins including those of hydrothermal origin. When a source of pure calcium carbonate is required it is normal to work vein calcite rather than limestone, a rock which commonly contains a proportion of other minerals. In the early days of the manufacture of polarizing microscopes it was usual to cut a rhombohedral crystal of pure transparent calcite (Iceland Spar) to make the Nicol Prisms. At the present time the artificial material polaroid is employed. Other uses of pure calcite include its employment as a decorative wall finish, as a soft ABRASIVE (tooth powders), and in the manufacture of whitewash and antacid tablets. Much pure calcium carbonate is manufactured from limestone.

The mineral aragonite is of the same chemical composition as calcite but it crystallizes in the orthorhombic system. Being less stable than calcite it is less abundant and it generally occurs only in relatively young rocks. It has no special value.

Calcium (Ca).   The fifth most abundant element in the earth's crust (3·6 per cent). Natural sources of calcium include LIMESTONE, DOLO-

MITE, MARL, shells and BRINES. Metallic calcium has limited metallurgical uses. Most calcium is used either in its original form or is heated to form the oxide, quicklime. The principal uses are in the production of CEMENT, as a flux in steel production and for agricultural purposes. Calcium products are high-volume, low-unit-cost commodities; their extraction mainly by quarrying can pose environmental problems. (◊FERTILIZER SOURCES.)

Caliper log. Diameters of wells are commonly measured throughout the depth of a well using a hole caliper. The caliper log is thus a vertical record of the well diameter, in the form of a continuous graph produced (by recording resistance changes) while the caliper is run up the well. Resultant data are useful for measuring diameters of old wells, locating cavity zones (which may represent horizons of increased permeability in the rock) and for supplying information on well CASING.

Cap rock. The covering or 'capping' of rock on top of an oil or gas reservoir or over a SALT dome. Usually it is an impermeable mass or layer of limestone, dolomite or gypsum a few metres to hundreds of metres thick.

Capillary fringe. The lowest subzone within the AERATION ZONE, occurring immediately above the WATER TABLE (Figure 6). It holds water above the ZONE OF SATURATION by capillary force acting against the force of gravity. The amount of water held in the capillary fringe depends upon the grain size of the material in which the water occurs. The capillary fringe in silt and clay may be more than 3 metres in thickness. In coarse sand or gravel, however, the fringe may have a thickness of a centimetre or less.

In ground water exploration the capillary fringe is essentially of academic significance except in a few cases where it shows the presence of the main body of water below the zone of aeration.

Capillary water. ◊SOIL WATER ZONE.

Carbon (C). The element upon which the chemistry of living things is based and the element consumed in the greatest quantities by the modern world. It is an essential component of all foodstuffs, FOSSIL FUELS and other sources of energy. Surprisingly perhaps, carbon forms only 0·02 per cent (by weight) of the earth's crust.

Carbonado. ◊DIAMOND.

Carbonates. ◊LIMESTONE.

Carnallite. ◊POTASH DEPOSITS.

Carnotite. ◊VANADIUM.

Carrolite. ◊COBALT.

Casing. Casing is installed in water wells for several reasons. It is used to seal off water-bearing formations penetrated by the well above the main aquifer, which may be contaminated or have undesirable characteristics. Casing is also used to prevent collapse of the well walls particularly near the ground surface where the material penetrated by the well may be deeply weathered and friable. The permanent well should have casing and a GROUT, as a sanitary precaution, to prevent movement of surface pollutants vertically downward to the intake portion of the well. Adequate depth of well casing is thus an important factor in sanitary protection. Steel pipe is used in the majority of wells but where corrosive waters may be encountered plastic tubes have to be used.

Cassiterite. A black or brown mineral, the dioxide of tin (SnO_2), usually showing an adamantine lustre and with a high SPECIFIC GRAVITY (7); the only commercial ore of tin. The principal reserves are the PLACER DEPOSITS of Asia (particularly Thailand) and Africa.

Catalytic clays. The CLAY MINERALS montmorillonite, halloysite and kaolin are employed after treatment as catalytic clays, principally in petroleum refining. The most suitable montmorillonitic clays are sub-BENTONITES with a low iron content. They are worked in Arizona and Mississippi.

Celestite. ⟡STRONTIUM.

Cement. Portland cement, so called because it was thought to resemble natural Jurassic Portland stone, is made from a mixture of about four parts limestone or chalk and one part of clay or mud. A small quantity of gypsum is usually added to act as a retarder during drying. China clay or bauxite are added to give high-alumina cement, which is more resistant to seawater and acid moorland waters and is also more REFRACTORY. Silica is added to increase resistance to alkaline soils.

The mixture should not contain more than 4 per cent magnesia (MgO), and it should also be free of uncombined quicklime, sulphur, iron pyrite and silica, unless the latter is specifically required.

Puzzuolan cement, which was manufactured by the Romans, is made from quicklime (CaO), sand and the volcanic ash of Puzzuoli. Cement-stone or cement rock is a naturally occurring argillaceous (clay-rich) limestone which can be used to make cement directly, or after the addition of a small quantity of extra limestone. MARL AND MARL-STONE may be used in the same way, but much additional limestone is required. Cement can also be manufactured from limestone plus pulverized fuel ash, and from oyster shells plus clay. Chalk is a favoured limestone for cement-making, because of its softness and ease of working. Some cement is also produced as a by-product of sulphuric acid manufacture from ANHYDRITE.

Cement works are generally sited at, or close to, a limestone quarry, and the clay is transported to the works. Many cement works in Britain are located close to the junction between the outcrops of a limestone and a clay or mud belt. In south-east England the works are generally situated on the chalk, close to workings either in the Gault clay or in estuarine muds, as for example in the Thames estuary region. Clay used for cement-making should be free from pyrite and gravel.

In Britain the principal limestones worked for cement-making are the chalk of Cretaceous age, Lower and Middle Jurassic limestones, and the Carboniferous limestone. Dolomitic limestones are unsuitable because of their high magnesia content. The principal clays or muds used are the Oxford and Kimmeridge clays of Jurassic age, the Gault clay of Cretaceous age, and alluvial and estuarine muds.

Most cement is used to make CONCRETE; a smaller proportion is mixed with sand to make mortar.

Ceramic clays. The principal raw material used in the ceramics industry. Each branch of the industry requires different types of clay according to the special properties desired. For example, STRUCTURAL CLAY PRODUCTS, such as bricks and tiles, are generally made from low-grade clay; REFRACTORY CLAY products from those containing a high proportion of the clay mineral, kaolin; kaolin is also a dominant mineral in CHINA CLAY, BALL CLAY and FIRECLAY.

Cerussite. Lead carbonate ($PbCO_3$) – one of the principal ores of lead. A soft (hardness 3–3·5), white or pale-coloured mineral with a high SPECIFIC GRAVITY (6·6) and an adamantine lustre. Cerussite commonly occurs with ANGLESITE in the zone of oxidation above veins of GALENA. It contains 78 per cent lead.

Cesium (Cs). A highly reactive element with extreme electropositive properties. It is used in the manufacture of electronic devices. The element occurs in the mineral, pollucite (hydrated cesium aluminosilicate), which is found in some granite PEGMATITES. Total world production (1968) was of the order of 20 tonnes.

Chalcedony. ⇨SILICA.

Chalcocite. A copper sulphide (Cu_2S) containing 80 per cent copper. Chalcocite is formed in the zone of SECONDARY ENRICHMENT by the alteration of other primary copper sulphides.

Chalcopyrite. A sulphide of copper and iron ($CuFeS_2$), containing 34·5 per cent copper, recognizable by its brass-yellow colour, softness (hardness 3·5–4) and greenish-black streak. Distinguished from PYRITE by its softness and from gold by its brittleness, chalcopyrite is the most important ore of copper.

Chalk. A soft, white, pure, fine-grained LIMESTONE of Upper Creta-ceous age, composed of very fine-grained CALCITE grains together with the remains of microscopic calcareous fossils. Beneath Denmark, much of northern France, and south and east England, there is a thick sequence of chalk. Its principal use is in the manufacture of CEMENT.

Elsewhere, other soft, friable limestones are often called chalk. These and other limestones are used for preparing whiting, a material employed in the making of paint, rubber and putty.

Blackboard 'chalk' is made from GYPSUM, not from chalk.

Chamosite. ⟡IRONSTONES.

Chert. ⟡SILICA.

Chile salpetre. ⟡NITRATE DEPOSITS.

China clay. Clay containing principally the CLAY MINERAL kaolin and which is suitable after treatment and blending for the manufacture of high-grade ceramic products, as opposed to STRUCTURAL CLAY PRODUCTS. Kaolin also has other important industrial uses.

The valuable properties of kaolin are its plasticity when moist, its ability to vitrify when heated to a high temperature, and its white or cream colour after VITRIFICATION. Clays for POTTERY AND STONE-WARE are not necessarily white burning. Calcite and iron pyrite in kaolin are detrimental to its use in ceramics, and the presence of quartz or other abrasive minerals render it less suitable for use in pigments and paper-making.

China clay was originally brought to Britain in the eighteenth century from the 'Kauling' (hence kaolin) range near King-te-chen in China. It was also worked from the Godolphin granite in Cornwall during the eighteenth century.

A few kaolin deposits are the result of the *in situ* weathering of rocks, such as granite, which are rich in feldspars. Many of the stratified deposits of kaolin are the result of the transportation of the kaolin grains in streams. Deposits of this type in Georgia and South Carolina occur in association with quartz sands in successions of Cretaceous age.

The famous high-grade kaolin deposits, those in the granites of south-west England, are the result of the pneumatolytic alteration of feldspars in granite intrusions. Although the St Austell region is the main producer, there is kaolin associated with the Land's End, Godolphin, Bodmin and Dartmoor granites. The St Austell deposits are worked using high-pressure water hoses to disaggregate the soft kaolin from the quartz and mica in the altered granite; the kaolin is separated by settling. The remaining quartz and mica present a serious disposal problem.

China stone or Cornish stone is hard, pneumatolytically altered granite within the St Austell intrusion, which when powdered is suitable for the manufacture of high-quality porcelain.

Elsewhere in the world kaolin deposits similar to those of the St Austell granite are worked in France (Limoges pottery), Czechoslovakia and China. China clay also occurs in Malaysia, Thailand, Tanzania and other parts of Africa.

Kaolin is also employed as a MINERAL FILLER in the manufacture of many products including paper, paint, textiles, rubber, plastics, some foods and cosmetics. Its use in paper-making is overwhelmingly the most important of these. Kaolin is added to CEMENT as a retarder and to increase its REFRACTORY and acid-resistant properties. It has also been used as a mild ABRASIVE and is employed in medicine. Production in 1971, in tonnes, was: USA, 4 500 000; UK, 2 736 000; India, 599 000.

Chloride. Occurs in great abundance in seawater where it averages about 19 000 ppm. The chloride content of rain in coastal areas is usually between 3 and 6 ppm diminishing rapidly inland to 1 ppm or less over a distance of about 150 kilometres. Most chloride in ground waters can be derived from a number of different sources: from ancient seawater (CONNATE WATER) entrapped in the sedimentary fabric of the rocks; from the solution of halite or other evaporite minerals (◊EVAPORITE DEPOSITS); it may also be concentrated by evaporation of rain or snow; or it may be introduced from surface pollutants.

Pumping from wells in COASTAL AQUIFERS or those adjacent to BRACKISH WATER inlets may induce the flow of saline water into the aquifer (Figure 3). Monitoring of the wells for increases in the chloride content of the ground water will provide good indication of possible contamination of the source.

Chloride, a negatively charged ion, can occur in natural water in concentrations from 0·1 ppm in polar snowfields to 150 000 ppm in brines. In ground water a chloride content above 250 ppm is regarded as objectionable for public water supply and more than 350 ppm is thought unsuitable for most irrigation and industrial uses. However, chloride concentrations above 1000 ppm are common in ground waters from arid regions where the water is used for domestic, agricultural and industrial purposes.

Chlorination. Used to disinfect water wells, killing off all potentially dangerous organisms that may be present. Whenever a new well is completed or an old well repaired there is a danger of contamination of the water supply source. Thus chlorine compounds are added to disinfect the water, the well thereafter being pumped to waste to remove all traces of the chlorine. Chlorination is by far the most common method of disinfecting water supplies.

Chlorite. ⬦CLAY MINERALS.

Chromite. A mineral – the only commercial source of CHROMIUM (Cr). A member of the SPINEL ((Mg,Fe,Zn,Mn)(Al,Cr)$_2$O$_4$) group of minerals. Chromite commonly occurs as dark grey, octahedral crystals found in basic and ultrabasic igneous rocks. The mineral contains 46 per cent chromium.

Chromium (Cr). An essential alloying element used in the iron and steel industry. The mineral CHROMITE is also used in the manufacture of REFRACTORY bricks and as a source of chromium compounds in the chemical industry.

South Africa, the USSR, Turkey, the Philippines and Rhodesia are major producers. World production of chromite ore (1974) was estimated at 7·3 million tonnes.

Chrysocolla. A hydrous silicate of copper (CuSiO$_3$ nH$_2$O) which is found, associated with other copper minerals, in the zone of oxidation.

Chrysotile. ⬦ASBESTOS.

Cinnabar. ⬦MERCURY.

CIPEC (Intergovernmental Council for Copper Exporting Countries). ⬦COPPER.

Clay. A name commonly applied to a fine-grained sedimentary rock which is plastic when moist and composed mostly of grains of less than 1/256 millimetre in diameter. The dominant minerals are those of the CLAY MINERAL group. Some clays also contain fine-grained calcite, iron pyrite, altered feldspar grains, muscovite flakes, iron oxides and organic material. A small proportion of silt- or sand-sized particles may also be present. The presence of organic material or iron pyrite commonly colours a clay black, ferric oxides give red or brown colours, while ferrous oxides may give a greenish colour and calcareous clays may be yellowish and pale-coloured.

The colloidal or near-colloidal particle size of clay minerals means that an aggregate of clay particles possesses an unusually large surface area, and when dispersed in water the mixture may be colloidal. Plastic clays are called 'fat', while those of low plasticity are called 'lean'. Fine-grained clays are generally more plastic. The 'green strength' of a clay is a measure of its ability to be handled when plastic; after drying its strength is measured in terms of 'dry strength'.

Consolidated clay (claystone) is one of the rocks belonging to the mudstone group (⬦MUD AND MUDSTONE). Clays or claystones are abundant in most parts of the world and in most rock sequences. Many clay rocks of Precambrian age have been converted to SLATE or SCHIST, as have others in the cores of fold-mountain belts. In Palaeozoic sequences most clays have been converted to claystones, while in Mesozoic and Cenzoic successions clays are found. This general increase of strength with geological age is not uniform everywhere.

Apart from clays which are the result of *in situ* hydrothermal alteration, or weathering, the majority were deposited following erosion and transportation as grains in water mainly little disturbed by strong currents or wave action. In both cases the source of the clay minerals is the chemical breakdown of unstable minerals such as the feldspars. Although many of the clays and claystones were laid down beneath the sea there are numerous examples of clays, deposited in former lake, river and glacial environments.

(◊ABRASIVES; ADSORBENT AND BLEACHING CLAYS; BALL CLAY; BONDING CLAYS; BRICK CLAYS; CATALYTIC CLAYS; CERAMIC CLAY; CHINA CLAY; CLAYS AS FILLERS; DRILLING MUD CLAYS; FIRECLAY; PIPE AND TILE CLAYS; PORCELAIN CLAYS; POTTERY AND STONEWARE CLAYS; REFRACTORY CLAYS; STRUCTURAL CLAY PRODUCTS; WHITEWARE CLAYS.)

Clay minerals. A group of related hydrous alumino-silicates, the principal constituents of CLAY and mudstone (◊MUD AND MUD-STONE). Individual clay mineral grains are generally platy or lath-like, and smaller than can be seen using a normal optical microscope. In order to see the grains clearly, an electron microscope is usually used, but the identification of the species of clay mineral is more commonly made with the aid of X-rays or by differential thermal analysis. Most clay minerals are crystalline layer or sheet silicates related to the micas.

Any amorphous substance in clays is commonly called allophane. Kaolinite, dickite and nacrite constitute the kaolin group, and are all hydrous alumino-silicates not containing significant quantities of other ions. Kaolinite occurs in both sedimentary and igneous rocks, while dickite and nacrite are restricted to hydrothermal deposits; kaolinite is the most abundant and important, being the principal constituent of CHINA CLAY. Halloysite has a composition similar to kaolinite but differs in structure and the amount of water of hydration. Montmorillonite possesses a structure in which ions of potassium, sodium, calcium, magnesium or iron may be exchanged, a property utilized in ADSORBENT AND BLEACHING CLAYS. Clays rich in montmorillonite are BENTONITES. Vermiculite possesses the unusual property of being capable of expanding to many times its original thickness when heated (◊EXPANDABLE ROCKS AND MINERALS). The illite group of mica-like (muscovite) clay minerals contains potassium ions, and although their structure is not unlike that of vermiculite they do not expand on heating. Illite is the most abundant clay mineral in many clays, and the dominant mineral in shale. Minerals of the chlorite group have a structure similar to the illites but they contain magnesium ions.

Chlorite occurs not only in clay rocks but also in phyllite, a metamorphic rock. Attapulgite and sepiolite, which resemble each other chemically in that they contain magnesium, are clay minerals which are not platy. Attapulgite is an important mineral in many adsorbent clays. Palygorskite is a clay mineral resembling attapulgite, and found in Eastern Europe.

Most clay minerals result from the chemical alteration during weathering and hydrothermal activity of pre-existing alumino-silicate minerals, especially feldspar. Some clay minerals grow in the sedimentary rocks in which they are found.

The properties of clay minerals vary from species to species. Thus it is necessary to investigate the mineralogy of a particular clay before its potential value can be assessed.

Clays as fillers. Clays of various types are used as MINERAL FILLERS in the manufacture of some types of paper, prints, plastics, reinforced roofing sheets, rubber, fertilizers, insecticides, textiles, inks and medicines. The employment of kaolin in the manufacture of some types of paper is a particularly important use of that mineral (▷CLAY MINERALS).

Claystone. ▷CLAY; MUD AND MUDSTONE.

Cleavage in rocks. A rock which is capable of being cleaved can be split into thin sheets between 1 millimetre and 3 centimetres in thickness. Where the planes of easy splitting are parallel to, and related to, the depositional layering of the rock, the structure is generally called lamination and the rock is said to be fissile. Shale is the fissile variety of mudstone. Where the cleavage planes are oblique to the beds or layers of a sedimentary succession, the structure is usually related to the development of FOLDS. Cleavage generally forms parallel, or nearly parallel, to the axial plane of the fold in which it is contained (Figure 12). Cleavage is best developed in originally fine-grained rocks such as mudstone or ash, although a cruder and generally more widely spaced cleavage is present in some folded limestones and sandstones. The principal processes involved in the development of cleavage are flattening, rotation of flaky grains such as micas and clay minerals, recrystallization, and solution transfer accompanying pressure. The most important form of cleavage is slaty cleavage, so named because it is the characteristic structure in the low-grade metamorphic rock called SLATE. When a thin, transparent slice of a slate is viewed under a high-powered microscope, it can be seen that the flaky grains of mica or clay mineral are arranged parallel to each other and the cleavage planes. There may also be a series of thin alternating dark- and light-coloured stripes of rock parallel to the cleavage. This feature results from processes which operate during solution transfer.

Another variety of cleavage is crenulation cleavage, a structure

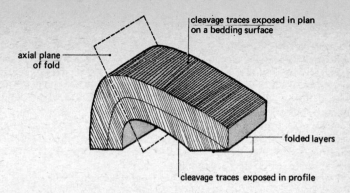

Figure 12. Cleavage in rocks.

comprising closely spaced parallel microfolds generally deforming an earlier slaty cleavage. Fracture cleavage is a structure formed by a set of unusually closely spaced failure surfaces. Equivalent structures to slaty cleavage in higher-grade metamorphic rocks are schistosity, named after the rock type SCHIST, and foliation, a banded structure characteristic of gneisses. Some rocks are cut by more than one set of cleavage planes indicating that they have been deformed more than once. Two sets of cleavage planes in a slate generally render it unsuitable for use.

Coal. The general term for rocks formed from the fossilized remains of freshwater plants in dense swamps or bogs. When the plants died they fell into the water where they were soon engulfed in the black, oxygen-deficient mud. Here they were protected from further decay and did not rot away. The mud itself was largely made of bacterially decomposed vegetable matter. Subsequent burial beneath sand or mud compressed and consolidated the planty layer, driving out water and other volatile substances. (The vegetable material underwent a series of changes as this proceeded.) Most of the world's coal beds were deposited millions of years ago in warm, humid regions. They have since been compressed by the weight of OVERBURDEN (1) or by earth movements to less than one-twentieth of the thickness of the original deposit. (◊ FOSSIL FUELS.)

Types of coal

From the point of view of value as FUEL, coals are classified according to the degree of change that they have undergone (Figure 13). With greater compression the vegetable material is reduced in volume, becomes blacker, harder, more brittle and the individual plant fragments become more difficult to distinguish. Such changes are said to indicate an increase in rank. In other terms, rank refers to the moisture,

53

volatile matter and fixed CARBON in the coal, and it increases with the proportions of fixed carbon present.

Peat is the lowest rank of vegetable deposit and is virtually uncompressed plant matter, brown to black in colour and spongy in texture. It may be formed in cool, swampy or boggy areas on land and made of mosses, grasses and small shrubs rather than large plants.

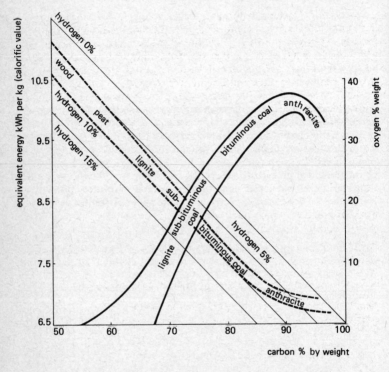

Figure 13. Different ranks of coal. The lignitic coals have a high proportion of oxygen and other volatile constituents, whereas the anthracite coals have a very high proportion of carbon. (After Open University, S26.)

Lignite or brown coal is low-rank coaly material with a high water and volatile content. It is typically a brown or black, crumbly or papery material, harder than peat but sometimes called a 'soft coal'.

Bituminous coal is a medium- to high-rank coal, typically compact, black and banded. Some layers are shiny; others dull and dusty. They contain less than 50 per cent of volatile constituents such as water and gases. Coals of lower rank are sub-bituminous. Many bituminous coals are also described as coking coals. That is, on heating in an oven they swell, give off water vapour, gas and tar, leaving behind coke.

Anthracite or 'hard coal' is a dense, black material, shiny and locally possessing an iridescence. These high-rank coals burn slowly, and contain more than 95 per cent of carbon.

Coals occur in rocks of Silurian age and younger. The world's most important coal reserves were formed in the Carboniferous and Permian periods. They occur in every continent. Most coal beds (seams or veins) are from a few centimetres to a metre thick, but rare exceptions reach several metres. Commercially worked seams are generally more than 50 centimetres thick and extend over scores of square kilometres.

The Chinese used coal 100 years or more BC. It was mined in Roman and medieval times, but its rise to prominence as an industrial fuel came with the Industrial Revolution.

All the major coal basins of the world have been discovered by now but not all are economically worth developing. The United States Geological Survey estimates that recoverable coal in seams more than 0·3 metres thick and at depths of less than 2000 metres amounts to about 8415×10^9 tonnes. The most efficient types of mining enable us to recover only about 50 per cent of known reserves. In 1970 the world production of coal and lignite was 2965×10^6 tonnes. (Figure 14.) (◊CONTINENTAL SHELF; OPENCAST (STRIP) MINING.)

D. G. Murchison and T. S. Westoll (eds.), *Coal and Coal-Bearing Strata*, Oliver and Boyd, 1968.
W. Francis, *Coal*, Edward Arnold, 2nd edn, 1961.
I. A. Williamson, *Coal Mining Geology*, Oxford University Press, 1967.

Coalfield. An area of sedimentary rocks which contains workable coal seams. Coals may crop out at the surface in these areas, as in Yorkshire, South Wales or Pennsylvania, or they may be completely cut off from the surface by overlying bodies of rock. Coalfields of the latter kind are known as concealed coalfields. The Kent coalfield is one such example, and a large concealed coalfield has recently been discovered south of Oxford.

Because many coalfields include rocks warped into a gentle, basin-like structure they are often referred to as coal basins.

Coalfields occur on all the continents; only in Antarctica are large coal resources as yet unworked. The northernmost coalfields in the world are worked by the Russians and Norwegians in Spitsbergen (latitude 78°N).

Coal measures. The coal seams of Europe, North America and some other parts of the world occur in thick sedimentary rocks in which a

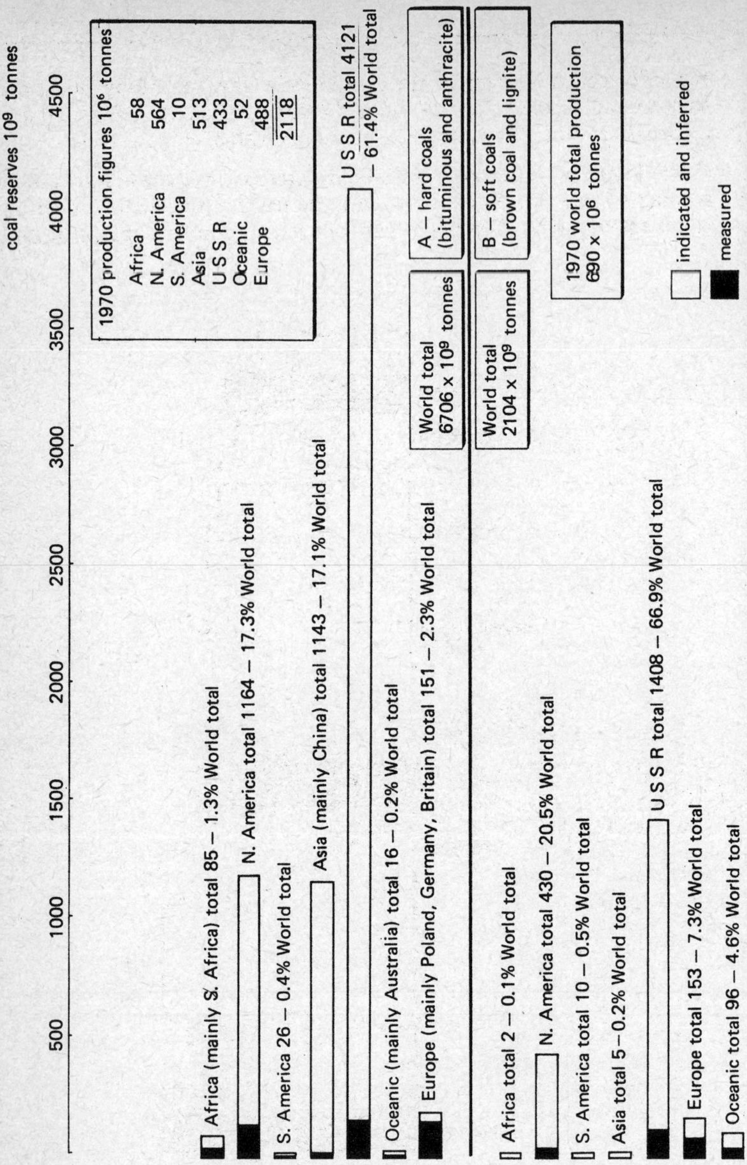

coal reserves 10⁹ tonnes

500 1000 1500 2000 2500 3000 3500 4000 4500

1970 production figures 10⁶ tonnes

Africa	58
N. America	564
S. America	10
Asia	513
U S S R	433
Oceanic	52
Europe	488
	2118

U S S R total 4121 – 61.4% World total

A – hard coals (bituminous and anthracite)

World total 6706 × 10⁹ tonnes

B – soft coals (brown coal and lignite)

World total 2104 × 10⁹ tonnes

1970 world total production 690 × 10⁶ tonnes

☐ indicated and inferred
■ measured

Africa (mainly S. Africa) total 85 – 1.3% World total
N. America total 1164 – 17.3% World total
S. America 26 – 0.4% World total
Asia (mainly China) total 1143 – 17.1% World total
Oceanic (mainly Australia) total 16 – 0.2% World total
Europe (mainly Poland, Germany, Britain) total 151 – 2.3% World total

Africa total 2 – 0.1% World total
N. America total 430 – 20.5% World total
S. America total 10 – 0.5% World total
Asia total 5 – 0.2% World total
U S S R total 1408 – 66.9% World total
Europe total 153 – 7.3% World total
Oceanic total 96 – 4.6% World total

Figure 14. World coal reserves by region. (After Open University, S26.)

56

repetitive rough layering of the components clay, sandstone, coal and shale occur. These rocks the miners called 'measures'. Many of the components other than coal have economic value. (Figure 15.)

Coastal aquifer. An AQUIFER which outcrops in a coastal area and is in HYDRAULIC CONTINUITY with the sea, e.g. the chalk aquifer of south-east England. The upper part of such an aquifer contains fresh,

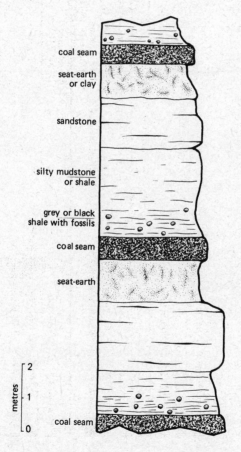

Figure 15. Coal measures: a vertical section as seen perhaps in a cleft, showing a typical coal measure 'cycle'. Several seams are present. Each coal seam is part of a separate cycle.

potable water while the lower part is saturated with saline water (Figure 16). Because of the difference in densities between freshwater and saltwater and because the WATER TABLE of freshwater is above sea level, the boundary between the two waters is in hydraulic balance (◊SALINE INTRUSION). In general terms the freshwater will extend to a depth about 40 times the height that the water table is encountered above mean sea level. This relationship is known as the Ghyben–Herzberg principle.

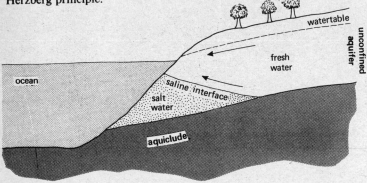

Figure 16. A coastal aquifer.

Cobalt (Co). Cobalt (crustal abundance 25 ppm) is used mainly as the metal in the production of high-temperature alloys (for gas turbine engines), in permanent magnet alloys and as cobalt-bonded, cemented carbides for high-speed cutting tools.

It occurs as a variety of sulphide, arsenides and sulpharsenides of which smaltite ($CoAs_3$), carrolite ($CuCO_2S_4$), skutterudite ($CoAs_3$) and cobaltite ($CoAsS$) are common. Cobalt minerals occur in complex ores associated especially with copper and nickel. All the cobalt of commerce is recovered as a by-product or co-product of other elements. (◊MANGANESE NODULES.)

Zaire is the major producer, accounting for about half the world's production (as a by-product of copper). Free-world production in 1974 (metal content) was estimated at 24550 tonnes.

Cobaltite. ◊COBALT.

Collector well. A collector or Ranney well comprises a central well with horizontal sections of perforated pipe, usually emplaced in a radial arrangement, to increase yield of water from the well. Frequently collector wells are located near to surface water such as a lake or river (Figure 17), and during pumping operations the collector well will lower the WATER TABLE and induce INFILTRATION from the surface water supply, through permeable material, to the well. This method of well supply is best in unconsolidated, permeable, alluvial aquifers where the resultant yield from the well may be of the order of several million gallons per day.

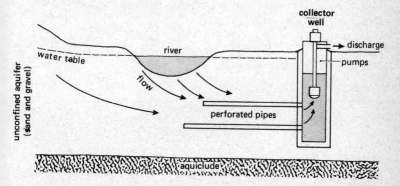

Figure 17. A collector well.

Collophane. ♢PHOSPHATE DEPOSITS.

Columbium (Ci). Columbium or niobium is used principally in various steel and other alloys. The source minerals are the complex oxides of the pyrochlore and columbite–tantalite series (♢TANTALUM). Brazil, Canada and the USSR are the major producers. Total free-world production (1973) of columbium concentrates was estimated at 13 000 tonnes.

Concrete. A strong and durable man-made rock made by mixing AGGREGATE with CEMENT and water. Gravel, crushed stone and sand may be used as aggregate. Concrete is today's principal constructional material.

Cone of depression. Cone of depression (or cone of influence) is formed in the lowering of the WATER TABLE or PIEZOMETRIC SURFACE by pumping water from a well (Figure 18). The shape and size of the cone of depression depends upon the pumping rate, the length of the pumping period, variations in aquifer permeability, the slope of the water table and recharge within the zone of influence of the well. (♢DEWATERING.)

Confined (artesian) aquifer. An AQUIFER covered by relatively impermeable material, and thus divorced from any atmospheric in-

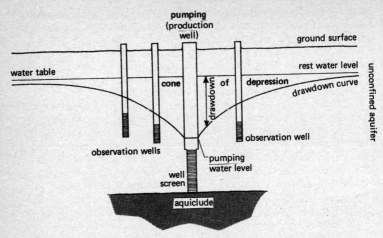

Figure 18. Definition of the cone of depression, drawdown and water levels surrounding a pumping (production) well in an unconfined aquifer.

fluences. Confined aquifers are saturated throughout their entire thickness. Most aquifers tend to pass downdip under a cover of relatively impermeable rocks (Figure 9), and thus unconfined conditions tend to give way to confined conditions in a downdip direction. In this state, because atmospheric influences have been removed, the ground water contained within the confined aquifer occurs within the pores of the aquifer at pressures greater than atmospheric. Thus when a well is drilled through the confining layer into the confined aquifer, ground water will rise in the well to a level above the top of the aquifer. This water level represents the artesian pressure in the aquifer, and is known as the PIEZOMETRIC SURFACE, a term only applied to ground water occurring under confined conditions. (◇ UNCONFINED AQUIFER.)

Conglomerate. A bedded, clastic sedimentary rock composed of granules, pebbles, cobbles or boulders of other rocks greater than 2 millimetres in diameter, and set in a matrix which is generally sand. The pebbles in a conglomerate are commonly of hard rocks such as granite or quartzite. If the pebbles are angular the rock is sometimes called a breccia.

Some well-consolidated conglomerates are used as sources of CRUSHED STONE, unconsolidated conglomerate is gravel.

Connate water. Water entrapped in the sediment fabric during the time of deposition of the sedimentary body. Thus it is water which has been out of contact with the atmosphere for a long time, commonly millions of years. The water might have originated from seawater or freshwater sources but when tapped by wells it, typically, is highly mineralized. Connate water, contrary to some definitions, is not only water which has virtually remained static since the burial of the surrounding sedimentary rocks; it can often be shown that connate water has undoubtedly migrated many miles since burial.

Connate water of the chloride type may be distinguished from other types of both surface and ground waters by chemical and other characteristics. The chemical composition of connate water reveals an enrichment in iodine, boron, silicon dioxide and combined nitrogen and calcium, and is low in sulphate and magnesium all relative to seawater. The isotopic composition of connate water shows $H^2/H^1 \leqslant$ seawater and $O^{18}/O^{16} >$ seawater. The temperature of connate water may vary between that of normal ground water (52°F) to slightly thermal. Connate water is itself of little economic importance but its presence has to be carefully monitored during basin-wide aquifer development.

Conservation. The exercise of some form of restraint and good husbandry in the use of a resource, so that it may be used to the best advantage. Where renewable RESOURCES are concerned, conservation aims to allow replenishment to keep pace with depletion so that a balance is maintained. Not only is the quantity of the resource to be considered but maintenance of quality is also important. Ground water is a resource in the management of which control of quality is very important. The conservation of non-renewable resources is a more difficult matter in some respects and may involve improvements in methods of resource extraction and utilization. Co-products and wastes may be reworked to ensure maximum yields. A further important aspect to be borne in mind is that the extraction or working of one resource should not be harmful to the utilization of another. In most mining areas, for example, ground water and surface water systems may be altered, and their quality may suffer, to be put right only at great expense.

As non-renewable resources are depleted, the need to exercise care in their use and to heed practices of conservation becomes more important. (◊ENERGY; FOSSIL FUELS; FUEL.)

Conservation of oil and gas
Many OILFIELDS have been tapped and in due course have virtually ceased to yield oil. Examples occur in California and Texas, USA, in the UK and Iraq. The conditions under which the oil was extracted were successful in producing only 30–35 per cent of the oil in the RESERVOIR ROCK; the remainder stuck to the sides of the pores and fissures in the rock. Methods of secondary recovery used today to obtain more of the oil from the pool (◊OIL POOL) include forcing the

oil out with gas, air or water or creating an underground fire to heat the oil, create more gas and increase the pressure to expel the oil.

Gas repressuring, i.e. the pushing of gas back into a reservoir to keep the oil moving, has been practised for many years. The process of pushing the gas back into the reservoir rock is now part of a pressure maintenance programme in many oilfields today.

Water flooding of the reservoir at a level below the bottom of the oil is an old and well-tried method which has on occasion led to the recovery of 25–30 per cent more of the oil than would have been produced with normal procedures.

Even the most successful production methods appear to leave about 30 per cent of the petroleum in the reservoir. Some wells may yield again after a 'resting time' which allows migration of gas or fluids at depth to restore enough pressure to allow production again.

Contact logging. A modified method of ELECTRICAL LOGGING. Contact logging involves measurement of the resistivity of both the fluids and materials occurring within uncased wells using electrodes which are lowered down the well and are pressed against the well wall using a rubber pad. As a part of the electric logging technique the contact log is important in ground water exploration work.

Continental shelf. Submerged extensions of the continental parts of the earth's crust, commonly lying at depths of less than 200 metres and largely made up of sedimentary rock. The continental shelf off the coast-line of Europe and most of the other continents is now the centre of attention for coal, oil and gas prospecting and extraction, gravel and mineral dredging, and mining beneath the sea floor.

Copper (Cu). A relatively scarce element (average 55 ppm in the crust) but, next to aluminium, the world's leading NON-FERROUS METAL. Free-world production (metal content) in 1974 was 6·2 million tonnes as compared with 11 million for aluminium. The metal is characterized by its high electrical and thermal conductivity, resistance to corrosion and its high strength. It readily alloys with other metals and is a component of a variety of brasses and bronzes. About half the copper produced is used as wire in the electrical industry. Other uses are diverse but include especially building construction (e.g. domestic water pipes, boilers), transportation, coinage and armaments.

Copper production has been threatened by substitution from other metals (especially aluminium), extreme price fluctuations and, in recent years, political intervention. The world's leading producers are the USA, USSR, Zambia and Chile. The latter two countries, which together with the Congo and Peru produce nearly a third of the world's copper, are members of the Intergovernmental Council of Copper Exporting Countries (CIPEC), an organization dedicated to establishing national interests over the influence of foreign mining companies.

Copper occurs in the native state and, because of its malleable and ductile nature, was one of the earliest metals used by primitive man. It combines with other elements to form a very large number of minerals, but only the sulphides – CHALCOCITE, CHALCOPYRITE, BORNITE; the carbonates – AZURITE and MALACHITE; and the silicate – CHRYSOCOLLA, are now of commercial importance. Rich hydrothermal vein deposits were the main source in the past but the bulk of present-day production comes from low-grade (averaging 0·6 per cent copper) disseminated ores. The working of such low-grade deposits is only economical if practised on a large scale. At the Bingham Canyon copper mine in Utah, for example, 96000 tonnes of ORE (1) and 225000 tonnes of OVERBURDEN (1) are extracted every day. Mining on such a scale presents serious environmental problems. (◊BACTERIAL EXTRACTION OF METALS; MANGANESE NODULES; PORPHYRY COPPER; STRATIFORM ORE DEPOSIT.)

Co-product. When two or more minerals or metals, of approximately the same value, are recovered from the same ore or mine they are termed co-products.

Coquina. ◊LIMESTONE.

Cordierite. ◊SILLIMANITE GROUP OF MINERALS.

Corundum. Ruby and sapphire are gem varieties of corundum (Al_2O_3) which in its common form is a valuable ABRASIVE. Corundum is the hardest mineral but one (hardness 9), and forms barrel-shaped crystals. It is concentrated in commercial quantities in some alumina-rich igneous rocks, in the margins of some pegmatites from which silica has been driven out by contact metasomatism (◊METAMORPH-ISM), and as a placer mineral in some so-called black sands (◊PLACER DEPOSIT).

Used loose, on paper or on cloth, corundum has been employed for optical polishing. It is also used in some impregnated cutting wheels. It is often employed in the natural mix of corundum, spinel and magnetite known as EMERY. Because demand outstrips supply, synthetic abrasives such as carborundum are often used.

Covellite. A copper sulphide (CuS) containing 66·4 per cent copper. It occurs in the zone of SECONDARY ENRICHMENT above primary copper sulphide minerals.

Crenulation cleavage. ◊CLEAVAGE IN ROCKS.

Crocidolite. ◊AMPHIBOLE; ASBESTOS.

Crushed stone. One of the most valuable earth resources for use in AGGREGATES despite its low monetary or unit value. Not only is it employed for making CONCRETE and ROAD METAL; it is also used for ballast, as GRANULES, as grit in agriculture, and as general fill.

The principal rock types worked for crushed stone are granite, syenite, diorite, gabbro, dolerite, rhyolite, trachyte, andesite, basalt, limestone, dolomite, sandstone, quartzite, schist, gneiss, amphibolite and marble. Although they vary in strength they are all relatively strong and uniform.

A common sequence of events during the production of crushed stone is churn-drilling, blasting, power-shovel loading, crushing, and finally screening and washing. The life of the jaws of a crusher is determined by the hardness of the rock being crushed. When granite is being crushed it may be a matter of only a few days, whereas when limestone is being crushed it may be several years.

Crust. ◊ THE EARTH.

Cryolite. A white or colourless mineral ($Na_3Al.F_6$), which is employed as a flux and solvent during the electrolytic manufacture of ALUMINIUM from bauxite. It is worked in significant quantities only near Ivigtut in Greenland, and occurs in association with siderite, fluorite and heavy-metal sulphides in a PEGMATITE. Cryolite is also known to occur in Russia, Spain, Canada and Colorado.

In addition to its use in aluminium refining, some cryolite is employed in the manufacture of glass enamels and glazes. Because the supply of natural cryolite is limited, synthetic cryolite is manufactured from hydrofluoric acid, sodium carbonate and aluminium.

Cryptomelane. A potassium-bearing manganese oxide of complex composition. It occurs as a weathering product of primary manganese minerals and as a constituent of some exhalative (hot spring) tufa deposits (◊ LIMESTONE).

Cupola. ◊ BATHOLITH.

Cut-off. The lowest GRADE at which an ore body is regarded as profitable. The cut-off value varies with, and depends on, the prevailing economic conditions.

D

Darcy's Law. Henry Darcy, a French hydraulic engineer, in 1856 carried out experiments with the flow of water through permeable material – sand – under differing heads of water. His conclusion, now known as Darcy's Law, was that the flow rate through porous media is proportional to the head loss and inversely proportional to the length of the flow path. This empirical law, more than any other contribution, provides the basis for present-day knowledge of ground water flow.

Desalination. The process whereby salt is removed from water to the extent that the dissolved minerals in the water are reduced to below 1000 ppm and preferably less than 500 ppm. Although desalting seawater to produce freshwater is carried out in a number of places, e.g. Saudi Arabia, it is frequently stated that desalination of mineralized water would be more economically viable for the desalting methods being tested. (⟡BRACKISH WATER.)

Descloizite. ⟡VANADIUM.

Detzeite. ⟡IODINE SOURCES.

Dewatering. A process normally employed in civil engineering works where a series of shallow, linked wells are used temporarily to dewater construction sites in wet ground. Foundation engineering schemes, sewage and water mains construction works often have to be carried out in saturated materials below the WATER TABLE. Using a well-point system and pumps, the normal water level can be pulled down, creating a composite CONE OF DEPRESSION, and the engineering works can be undertaken in 'dry' ground. Proper dewatering should also prevent the likelihood of upwelling saturated sand, etc., developing in the bottom of an excavation.

The term dewatering is also employed in both coal and metalliferous mining areas where the workings are continuously pumped in order that the winning of coal and metal ores can be carried out under dry conditions.

Dew point. The point at which the formation of dew starts for given temperature and humidity conditions. Thus if air is cooled, without barometric pressure changes, until it becomes saturated, the corresponding temperature is called the dew point. If the dew point is above 32°F, condensation will be in the form of dew; whereas if it is below 32°F, frost will be formed.

Diabase. ⟡DOLERITE.

Diagenesis. The process of consolidating a loose sediment such as sand into a solid cohesive rock, involving the cementation of the particles together.

Diamond. A cubic variety of crystalline CARBON (C), and the hardest known mineral. Crystals are commonly octahedral in shape and have a perfect octahedral cleavage. Uncut stones have a greasy appearance. They are usually colourless or pale yellow but may be other colours. The unit of weight is the carat which is 200 milligrammes; a one-carat diamond is about a quarter of a inch across. Famous diamonds include the Cullinan (3106 carats) found at Pretoria in 1905, the Jonker (726 carats) found in 1934 and said to be of the finest quality, and the Star of Sierra Leone (969·8 carats) found in 1972.

Although diamonds are usually thought of as gemstones, over 80 per cent of all diamonds mined are used for industrial purposes. Bort (colourless fragments), carbonado (coke-like) and ballas (radiating aggregates) are varieties of industrial diamond. The principal uses are as ABRASIVES for cutting and grinding tools. Diamond is artificially synthesized from carbon at elevated temperatures and ultra-high pressures; over 10 million carats of synthetic diamond are produced annually.

The primary source of natural diamond is in a volcanic breccia known as Kimberlite but most diamonds are recovered from PLACER DEPOSITS. Major producers are Zaire, the USSR, South Africa, Ghana and Angola. Total world production (1974) was estimated at 50·4 million carats.

Diaspore. ⟡BAUXITE.

Diatomaceous earth. ⟡DIATOMITE.

Diatomite. A siliceous sedimentary rock consisting of diatoms, microscopic planktonic algal skeletons or shells composed of hydrous opaline SILICA. Each shell has a very large surface area compared with its size.

Deposits of diatomite, also known as diatomaceous earth (tripolite) or kieselgur, are generally soft, friable, porous and of low density. They can absorb up to three times their own weight of water. When pure they are white, but impurities can colour them grey, brown or black. The majority of commercial deposits occur in rocks of Cainozoic age, especially those of Neogene times. They accumulate where there is a plentiful supply of silica in solution, as for example in active volcanic regions, or in areas where volcanic rocks are undergoing weathering. Many diatomite deposits were laid down in freshwater environments such as lakes. Diatomite also accumulates in some marine environments, especially where the salinity is low and the water is cold.

Diatomite is employed as a lightweight AGGREGATE, for industrial filtration, as a MINERAL FILLER, as an ABRASIVE (for example, in car polishes and on match boxes), in sound and thermal insulation, and as an absorbent (⊘ABSORPTION ON MINERALS) and catalyst. Formerly it was used in the manufacture of dynamite.

Dickite. ⊘CLAY MINERALS.

Dimension stone. Stone which is marketed in blocks or slabs of a specified size. Much of it is used for building, either for structural purposes, or for cladding, facing or flagging. The minerals in dimension stones should be stable in humid atmospheres, and the rock should be crystalline or well consolidated. Individual blocks should be free of weakness planes, and if the stone is to be used for structural purposes it should be strong. The principal rock types used as dimension stone are granite, diorite, syenite, sandstone, limestone, some ironstones, gneiss, slate and serpentine.

When a rock mass is being considered for working as a dimension stone, it is important that a sufficiently plentiful supply of material of the same colour and texture is available, and that the stone can be worked with the minimum of blasting, a process that is liable to introduce cracks into otherwise intact blocks.

Stone to be used for statuary purposes should generally be attractive, not too hard, even-grained and workable by hand. Marble is probably the best-known stone used by sculptors.

Some porous stones, such as the Middle Jurassic oolitic limestones of the Cotswold Hills in England, have been widely used for building. Despite their attractiveness they are liable to chemical decomposition; in some cases this effect can be minimized by applying an outer protective coating. (⊘ARTIFICIAL STONE.)

Diopside. ⊘PYROXENE.

Diorite. A coarse-grained, plutonic igneous rock of intermediate composition, commonly containing the minerals plagioclase feldspar, hornblende and biotite. Some diorites also contain a small percentage of quartz, and others contain some augite. Diorites are often found in stocks or bosses or in parts of a larger BATHOLITH. Some diorites have been worked for DIMENSION STONE but most are of value only for CRUSHED STONE.

Dip. The dip of a rock layer or other surface, such as a fault, in a rock mass is the maximum angle that the layer or surface makes with the horizontal (Figure 19). The direction of dip is the direction measured in a horizontal plane of the maximum inclination, and it is always at right angles to the direction of STRIKE.

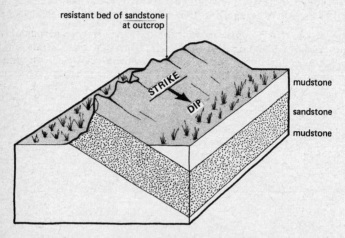

resistant bed of sandstone at outcrop

STRIKE

DIP

mudstone

sandstone

mudstone

Figure 19. Outcrop of a tilted bed of sandstone to illustrate its dip and strike.

Discharge curves. In any analysis of river or stream flow, discharge measurements can be made by dividing the stream cross-section into a number of parts for each of which the area, velocity and discharge are determined separately. Addition of these partial discharges will result in a total discharge value for the stream. A discharge curve can be produced after such measurements have been plotted against the corresponding stream gauge heights. After this calibration has been made, the gauge readings can be used directly to determine the river discharge.

Divining rod. A fork-like branch or wire normally used in DOWSING or water-witching, i.e. locating subsurface water supplies.

Dolerite. A dark-coloured, medium-grained, hypabyssal igneous rock which characteristically occurs in DYKES or SILLS. One well-known sill is the Whin Sill of northern England. In the region where it is worked it is known as Whinstone. Dolerites contain calcic-plagioclase feldspar, augite or related pyroxenes; some contain olivine, and a few contain free quartz. They are equivalent in composition to BASALT and

GABBRO. In North America dolerite is called diabase, a name which in Britain is generally reserved for metamorphosed dolerites.

Many dolerites are cut by closely spaced joints which generally make them of little value as DIMENSION STONE. They are however widely quarried for CRUSHED STONE, and they make excellent ROAD METAL. Where a relatively thin, vertical dyke of dolerite is quarried, the workings are of necessity long, narrow, trench-like excavations. ($\Diamond$GRANULES.)

Dolomite. The trigonal mineral which is the double carbonate of magnesium and calcium $((CaMg)CO_3)$. When pure this soft mineral is white and commonly occurs as rhombohedral crystals with gently curved faces. The name is also used for the carbonate sedimentary rock which is predominantly composed of the mineral dolomite. LIMESTONES containing both CALCITE and dolomite are called dolomitic limestones, those with some magnesium in the lattices of the calcite grains are commonly known as magnesian limestones.

Although the mineral dolomite is precipitated as a primary sedimentary mineral, especially during the formation of EVAPORITE DEPOSITS, many dolomites and dolomitic limestones are the result of the late alteration of some, or all, of the calcite in a limestone, a process known as dolomitization. It commonly involves a significant decrease in volume which may lead to the development of additional pore spaces and VOIDS in the rock. These may act as reservoirs for oil or water. Dolomite also occurs as a hydrothermal vein mineral.

Well-known dolomites or dolomitic limestones in Britain are: the Durness limestone of Cambro-Ordovician age in north-west Scotland; some of the limestones in the Carboniferous limestone series of the southern Pennines, South Wales and the West Country; and the magnesian limestone of Permian age, the outcrop of which stretches from Nottinghamshire to the Durham coast.

Dolomites and dolomitic limestones are worked for many purposes. Some are quarried and washed for use as AGGREGATE. The Permian magnesian limestone near Mansfield in Nottinghamshire has been worked for DIMENSION STONE. Other uses include the extraction of MAGNESIUM or MAGNESIUM SALTS, employment as a REFRACTORY material (less satisfactory than MAGNESITE but commoner and cheaper), and the manufacture of carbon dioxide. Used as a FERTILIZER, the magnesium content of dolomites prevents the staggers in cattle. Some dolomites are employed in the manufacture of GLASS, ceramics and rock wool ($\Diamond$MINERAL AND ROCK WOOL), as an ABRASIVE ('Vienna Lime') and as a metallurgical flux. ($\Diamond$CRUSHED STONE; GRANULES.)

Dowsing. The method of prospecting for water with a DIVINING ROD (also known as water-witching). The method lacks scientific justification and successes are ascribed to the psychic powers of the person using the divining rod. The forked implement is held in both hands while traversing an area until the butt end of the rod is attracted downward 'by the subsurface water'. Divining rods are not only used in locating water but diviners have been called upon to locate ores, criminals, Protestants and lost animals! The method is therefore one to which much scepticism has been directed.

Drainage network. Relates to a system of rivers or streams draining a particular basin. The characteristics of a drainage network may be described by a number of factors including the order of streams, length of tributaries and the stream density.

Drawdown. When a water well is pumped, ground water is removed from the aquifer surrounding the well and the natural water level – either the WATER TABLE or PIEZOMETRIC SURFACE depending on the aquifer condition (unconfined or confined) – is lowered. This difference in water level is known as the drawdown (Figure 18) and is measured in metres or feet as the distance between the natural water level of the aquifer and the pumping water level in the well. The drawdown thus represents the head, in metres or feet of water, that causes water to flow through the aquifer towards the well at the rate that water is being abstracted from the aquifer. A resulting drawdown curve can be constructed which represents a plot of drawdown with radial distance from the pumping well. It will thus define part of the CONE OF DEPRESSION around the well.

Dredging. Originally the operation of pulling an underwater device over an underwater area to be excavated. Today dredging may be carried out to develop or maintain water depth for navigation, to transfer earth from beneath water on to the adjacent land or to recover underwater deposits or marine life. Dredging was used by the ancient Assyrians and Chinese and by the Italians and Dutch since the fifteenth century. Today it is a highly developed technology in which material is raised either by mechanical means or by pumping to a floating platform or ship. The basic types of dredge may be described as:

1. A scraper (or dragline) where a scoop is drawn over the bottom by a line.
2. A dipper – essentially a scoop on the end of a boom.
3. A grab where a mechanical grab bucket is raised and lowered.
4. A ladder or bucket chain where an endless belt of buckets is lowered to dig into the substrate and carry it to the surface.
5. Hydraulic dredges which pump water and sediment to the surface to be sieved. A concentration of 20 per cent of solid materials may be raised by this method.

Gold and metallic ores have been dredged on a large scale from rivers and swamplands in North America, Australia and Asia, but the principal field of interest today is the recovery of minerals from the CONTINENTAL SHELVES of the world. Gravel and other AGGREGATES are dredged in great bulk around Europe. Further development of dredging techniques is planned to recover MANGANESE NODULES from the sea floor at greater depths, but so far the cost of such operations appears to make them uneconomic. (Figure 20.)

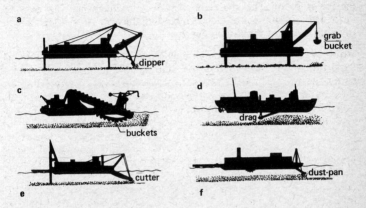

Figure 20. Dredging: (a) dipper dredge in position with dipper ready to excavate; (b) grab dredge with bucket raised; (c) bucket-ladder dredge with ladder lowered to excavating position; (d) suction dredger with drag lowered for raising material; (e) cutter-head hydraulic dredge with cutter lowered for excavating; (f) dustpan hydraulic dredge in operating position. Dredged material may be stored in hoppers on board, transferred to tenders, or removed by floating pipelines as in (e) and (f). (After *Encyclopaedia Britannica*.)

Drilling methods. Most large-diameter, deep, high-yielding water wells and all oil or gas wells are made by drilling (Figure 21). The choice of drilling equipment depends on the use to be made of the equipment as well as economic and geological factors. Geological conditions normally lead to two basic approaches to well construction. A well that penetrates a consolidated (bedrock) aquifer will consist basically of a cased portion, usually extending through the upper, more friable and weathered materials, and an open, uncased hole in the rock below. A well that taps a saturated aquifer comprising unconsolidated sand and/or gravel must be provided with a well CASING above the WATER TABLE and a well screen (slotted casing) in the ZONE OF SATURATION.

71

Three main methods of drilling and well construction are usually employed – cable-tool (percussion), hydraulic rotary and reverse rotary methods. These methods may individually be suited to one particular aquifer condition and not to another.

The cable-tool method is one of the oldest and still one of the most popular methods of well drilling. Simply, the method involves the

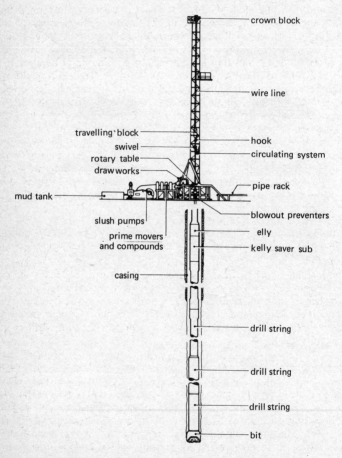

Figure 21. Most wells for water, oil or gas and boreholes to obtain samples are drilled by rotary methods. The essential parts of a rotary rig are shown above. They are not all shown to the same scale.

lifting and dropping of a string of drilling tools in the hole. The drill BIT breaks or crushes hard rock into small fragments or, when employed in unconsolidated aquifers, the drill bit loosens the material. It is in this latter situation that this method is least effective. In addition to a drilling rig and the percussion tools a bailer is employed to remove the drill cuttings from the well. Drilling rates can vary between 2 and 35 metres per day depending upon the material being drilled.

Rotary drilling methods have become increasingly more important since the 1920s. This is because of the greater drilling speed of this method and because well casing is normally not required during drilling. Hydraulic rotary drilling consists of the cutting of a hole by means of a rotating bit and removal of the cuttings by continuous circulation of a drilling fluid as the bit penetrates the aquifer. The drilling fluid is commonly a clay/water mixture, and the cuttings are carried up the hole by the rising drilling mud. The speed of rotary drilling again depends upon geological factors and may range between 10 metres per day for dense, consolidated rocks to 100 metres or more per day in soft, unconsolidated sediments. Reversed-circulation rotary drilling is undertaken with the flow of drilling fluid reversed as compared with the conventional rotary system. Here the drilling fluid, commonly water, and its contained cuttings are removed upward inside the drill pipe by a suction process and are discharged by the pump into a settling pit adjacent to the rig. These rotary methods can be used in the construction of wells up to 1·25 metres in diameter.

Some of the greatest problems involved in rotary drilling are encountered in aquifers with highly permeable zones such as cavernous limestone and basalt. The main difficulty is caused by the loss of drilling fluid into the highly permeable horizons. This can result in collapse of the well walls and loss of expensive drill bits and rods.

Other drilling methods sometimes employed include hand drilling (augering), jetting and AIR DRILLING. The last is best suited to impermeable rock material where the cuttings produced by drilling are relatively small.

Drilling mud clays. During the rotary drilling (◊DRILLING METHODS) of deep oil wells, drilling mud is circulated in the hole to remove drill cuttings, coat the hole wall, cool and lubricate the BIT, and prevent blow-outs. Clay is added to water to increase its viscosity and in many instances to provide a thixotropic material (◊THIXO-TROPY). The most suitable clays for this last purpose are swelling BENTONITES. In areas of saline ground waters, non-swelling bentonites, or the clay mineral attapulgite, are used, but their thixotropic properties are limited.

To prevent blow-outs, the density of drilling mud is increased by the addition of powdered dense minerals such as barite.

Eh (redox potential). Eh is a measure of the energy required to remove electrons from ions in a given chemical environment. The solubility of certain elements will depend upon their oxidation state which is determined by the redox potential (oxidation-reduction potential) or Eh of the environment.

Electrical logging. In the subsurface geophysical investigation of water wells it is common today to undertake an electrical survey. It serves to verify and supplement the descriptive logging of the hole which the driller notes as drilling proceeds. Electrical logging (often regarded as synonymous with resistivity logging and potential logging) consists of recording the apparent resistivities of the subsurface rock strata and the spontaneous potentials generated in the well, both plotted in terms of depth below ground surface. These two properties are directly related to the characteristics of the subsurface formations and to the quality of water within them.

These electrical properties can only be determined in fluid-filled, uncased wells when current and potential electrodes can be lowered to measure the variation in electrical characteristics of the subsurface media with depth.

Resistivity logs, using artificially induced electrical currents, and the resultant resistivity curves, indicate the kind of material penetrated by the well and enable freshwater and saltwater to be distinguished. In abandoned wells the logs can show the exact location of casing in the wall. Resistivity of unconsolidated, porous materials is controlled by the porosity, packing, water resistivity, degree of saturation and temperature. Clays and saltwater sand bodies give low resistivity values whereas freshwater sands give moderate to high values and consolidated sandstones and limestones high values. The resistivity of a ground water body will depend upon the IONIC CONCENTRATION of that body and the ionic mobility of the salt solution. The electrical conductivity (conductance), however, is the ability of water to conduct an electrical current. It is thus the reciprocal of resistivity, and also will depend upon the temperature, type of ions present in the water and the concentration of ions. Rainwater, with a low ionic concentration, is a poor conductor of electricity and has a high resistivity. Conversely, saline water is a good conductor with a low resistivity.

Potential logging methods are used to measure the natural electrical potentials (currents) within the ground. These potentials are referred to as spontaneous, or self, potentials or as 'S.P.'. The S.P. curve is an integral part of the complete electric log. (◊CONTACT LOGGING; GAMMA LOGS.)

Eluvial deposit. Decomposed material on hillsides or valley slopes containing a mineral(s) resistant to WEATHERING, e.g. gold or tin. The resistant mineral is still near its place of origin but may have been concentrated because of selective chemical weathering, and removal in solution, of part of the original host rock.

Emery. A natural mix of CORUNDUM, plus magnetite, HEMATITE or SPINEL. Emery is a widely used hard ABRASIVE; its cutting quality depends upon the corundum content but the presence of the other minerals makes it less harsh.

The principal occurrence of emery is in metasomatically altered (⬦METAMORPHISM) limestones and basic and ultrabasic igneous rocks which have been converted to chlorite-hornblende schists. The best-known deposit of this kind occurs on the Greek island of Naxos. Other deposits of emery occur in Turkey, New York State and the Urals. Some emery also occurs in residual deposits which are derived from the weathering of primary deposits.

Emery is used in grinding wheels, emery cloth, for glass polishing, and for incorporating in non-slip, wear-resisting concrete floors. Production in 1971, in tonnes, was: Turkey, 83 100; Greece, 7000.

Energy. The physicist defines energy as the ability to do work, kinetic energy being the energy of motion (wind or running water) and potential energy being the energy of position. To mankind energy is next in importance to our food and water as an essential resource. It is necessary in every field of human activity and has been derived from fuels, the wind, flowing water and of course from the toil of men and animals. (Figures 24, 25.)

The consumption of energy by nations is commonly reckoned in kilowatt-hours per year and the figures are in the million million order of magnitude for large industrial nations. In the United States, the

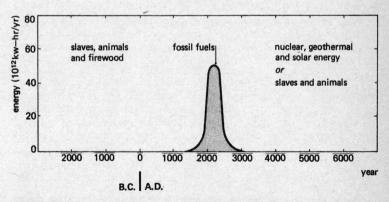

Figure 24. The energy sources of mankind, illustrating the brief duration of the fossil-fuel epoch. (After Menard.)

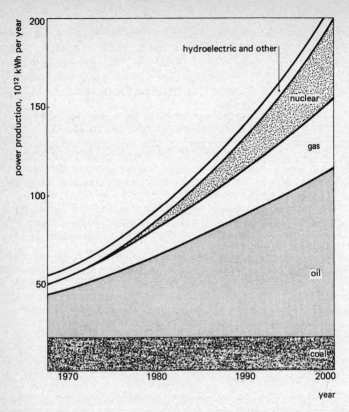

Figure 25. Predicted world power production. (After Open University S26.)

world's biggest consumer, the rate of energy consumption could be expected to double in less than thirteen years. World consumption is increasing even faster. New sources of fuel and energy are necessary in very large measure to meet this exponential growth. Developed countries use over 85 per cent of the world's energy. The USA with about 6 per cent of world population uses more than a third of all consumed energy.

Broadly speaking, 40 per cent of energy used comes from oil and about the same from COAL: nearly 20 per cent is derived from NATURAL GAS. Very little is derived from other sources and it is to these other sources that man must turn. Coal, oil and gas are non-renewable, solar energy, wind and water power are virtually renewable

78

for ever, but efficient technologies have yet to be developed more extensively to make any significant use of them. (⋄FOSSIL FUELS; FUEL; GEOTHERMAL ENERGY; PETROLEUM; PRIMARY ENERGY SOURCES.)

E. Cook, The Flow of Energy in an Industrial Society, *Scientific American*, vol. 224 (1971), pp. 134–47.
M. K. Hubbert, The Energy Resources of the Earth, ibid., pp. 60–87.
D. Luten, The Economic Geography of Energy, ibid., pp. 165–75.
C. Starr, Energy and Power, ibid., pp. 36–49.
G. Holister and A. Porteous, *The Environment: A Dictionary of the World Around Us*, Arrow Books, 1976.

Ephemeral streams. Streams which carry only surface RUNOFF and hence flow only during and immediately after a rainfall period or snow-melt period. They have no well-defined channels, and the drainage basin in which these streams occur comprises either relatively impermeable material or material in which the WATER TABLE is well below the stream bed throughout its entire length.

Epigenetic deposit. A mineral deposit that has formed later than the enclosing or associated rocks. The opposite of SYNGENETIC DEPOSIT.

Epsom salt (epsomite). ⋄MAGNESIUM SALTS.

ERTS (Earth Resources Technology Satellite). Satellites originating from America's NASA laboratories, which are designed to monitor many aspects of the earth's surface as it orbits the planet. It is intended to transmit photographic images to earth which will be useful in providing data about geological formations, water, soils and vegetation. The satellite ERTS 1 was launched from California on 23 July 1972 into polar orbit about 560 miles above the earth; it was designed to cover the entire earth's surface at least once every eighteen days. It completes an orbit in 103 minutes. Although intended to last only one year its work is still continuing (February 1976). The satellite carries three cameras, a multi-spectral scanner and a data-collection system. The cameras failed soon after the launching but the scanner has provided extremely valuable, high-quality images of the earth's surface.

Data are transmitted to receiving stations in Alaska, Maryland, Canada and Brazil partly in the form of live-scan (TV-type) pictures. Uncharted lakes in Bolivia, deeply buried faults in the United States, possible petroleum areas in Alaska and areas of potential ground water value in Kenya have all been located by ERTS 1 early in its programme. A second satellite is now in orbit and will take on the functions of the first as the original equipment fails. The satellites are now referred to as LANDSAT 1 and LANDSAT 2 (⋄LANDSAT). (⋄⋄REMOTE SENSING.)

79

Evaporation. The change of water from liquid or solid state into a gaseous state through heat exchange; an integral part of the HYDRO-LOGICAL CYCLE (Figure 3). Evaporation can take place from free-water surfaces within the atmosphere, or from surface-water films on soil particles or vegetation, or from water surfaces, such as reservoirs, rivers and oceans. With vegetation the evaporation is from leaf surfaces and thus the amount of evaporation will vary with the type of plant and the surface area of individual leaves. The origin of this water film on plants is the process of TRANSPIRATION, but the movement of this water back into the atmosphere is evaporation. Sometimes the term evapotranspiration is used to describe these mutually involved processes.

Evaporite deposits. Composed of concentrations of minerals which have been crystallized as a consequence of the evaporation of bodies of highly saline or alkaline water. Seawater which has become separated, or partly separated, in gulfs or lagoons from the main body of oceanic water evaporates in hot, arid climates provided that there is no supply of freshwater to the lagoon. Evaporites are also deposited in some ALKALI and SALINE LAKES which are basins of internal drainage in arid areas. Many of these lakes are seasonal or playa lakes.

Ancient evaporite deposits are relatively abundant and widely spread in areas of rocks younger than the Precambrian.

The mineral species present in a particular evaporite deposit reflect the chemistry of the parent waters. Not surprisingly the suite of minerals associated with former bodies of seawater is different from that associated with former lakes which were not connected to the sea.

During the continuous evaporation of a body of seawater the evaporite minerals and rocks are precipitated in a definite order: limestone and dolomite, gypsum, anhydrite, salt and finally, if drying up is complete, the minerals of potash deposits and magnesium salts. Variations in the order of crystallization may be caused by temperature changes and the introduction of fresh seawater. Layers of evaporite minerals are generally interbedded, commonly in 'red-bed' sequences, with limestones, sandstones and mudstones.

In order that tens or hundreds of metres of evaporites may accumulate it is necessary to suppose that the lagoon or gulf was recharged with seawater from time to time. Seawater contains only a very small fraction of dissolved solids, insufficient in a shallow sea to account for the great thicknesses of evaporites which may accumulate in a relatively short interval of geological time. A favoured idea to explain how this 'topping up' might happen is the bar theory. It is imagined that the lagoon or gulf is separated from the open sea by a bar or sill, at about sea level, which prevents the outflow of the denser saline water. Its height is, however, insufficient to prevent new seawater entering the basin at times of exceptionally high tides. A present-day example which approaches this model is the Gulf of Karabugaz, off the Caspian Sea. Evaporites in the sediments beneath the floor of the Mediterranean are thought to have formed when that ocean dried up during the late Tertiary.

The mineralogy of the evaporites of saline and alkaline lakes which were not connected to the sea depends upon the chemistry of the rocks over which the stream supplying the lakes have flowed. They are generally richer in carbonates, sulphates or borates than marine evaporites.

Other evaporites, particularly the nitrates, are concentrated in superficial deposits as a result of efflorescence and the evaporation of rising ground water.

Evapotranspiration. ◊EVAPORATION; TRANSPIRATION.

Expandable rocks and minerals. The low density, low unit value, and the thermal and acoustic insulation properties of this category of lightweight AGGREGATES makes them much in demand. The three principal raw materials in the group are bloatable clays, perlite and vermiculite. When they are rapidly heated they bloat or exfoliate because gas, mainly in the form of water vapour, is driven off as the material vitrifies (◊VITRIFICATION).

Suitable clays, mudstones and slates for bloating have a low alkali content. The only practicable method of determining whether a clay is suitable for bloating is to treat it in a kiln. Lightweight aggregates from this source have a great potential because the raw material is so widespread. They are used for making lightweight concrete slabs which have the same strength as conventional concrete but are 30–40 per cent lighter.

Perlite is an acid, glassy, spherulitic, volcanic igneous rock containing dissolved water, which, when the rock is crushed and then heated into its softening range, results in an expansion of 4–20 times the original volume of the rock. Some andesites and basaltic ignimbrites also behave in this way. Devitrification generally renders the rock less suitable for bloating, hence lavas older than the Cretaceous are rarely used. The principal uses of perlite are as a substitute for sand in the manufacture of GYPSUM plaster, in the production of lightweight CONCRETE AGGREGATES, oil-well CEMENT, soil conditioner, ABRASIVES, some plastics, rubber, paper and textiles, and as a loose fill insulator. (◊MINERAL FILLERS.)

Vermiculite, a member of the CLAY MINERAL group, is an inelastic, mica-like, complex hydrated magnesium, calcium, iron, aluminosilicate. There are water molecules between the silicate layers in its structure. When the water is driven off, vermiculite may expand to 22 times its original thickness.

Biotite and phlogopite may be altered to vermiculite by weathering or hydrothermal solutions. Most deposits are associated with basic and ultrabasic igneous rocks and with some high-grade mica schists. It may also occur in the clay component of some soils and in association with carbonatites and metamorphosed limestones.

The principal uses of vermiculite are in the manufacture of lightweight plaster AGGREGATE, insulating CONCRETE and loose-fill insulator. It is also employed as a soil conditioner, as an absorbent material, in powder fire extinguishers, in DRILLING MUD, oil-less lubricants, and as a filler in plastics and rubber (◊MINERAL FILLERS).

Production in 1971, in tonnes, was: for vermiculite, USA, 273900; South Africa, 132070; for perlite, USA, 450000; Greece, 160614.

Extrusive rocks. ◊IGNEOUS ROCKS.

F

Fault. A fault in a rock mass is a FRACTURE surface along which there has been permanent displacement (Figure 26). Where bedded rocks are disrupted by a fault it is not always easy to trace any particular layer across the fault. The amount of displacement on faults varies from a few millimetres to hundreds of kilometres. A fault zone is a relatively narrow zone containing several closely spaced fault planes and other structures such as JOINTS developed during faulting. A fault plane may bear striations which indicate the sense of shear on the fault, and it may be accompanied by fault breccia or by gouge. Fault breccia is a rock composed of angular fragments which are commonly supposed to be the result of friction during faulting. Gouge is a clay-like rock composed of clay or crushed rock debris.

There are three principal categories of faults: normal faults which result when part of the earth's crust is being stretched; thrusts or reverse faults which are a consequence of lateral compression; and wrench or strike-slip faults, which are formed when rocks are being compressed laterally in one direction and being elongated laterally in another. Many thrusts are intimately associated with FOLDS, structures which are also a consequence of lateral compression.

To the economic geologist, faults are of concern because they may disrupt mineral deposits and coal seams. Faults may act as channels or barriers to ground water or oil flow according to whether they are accompanied by fault breccia or gouge.

Fayalite. ⟡OLIVINE.

Feldspar. The family of minerals which includes orthoclase and microcline, both of which are potassium alumino-silicates ($KAlSi_3O_8$), the former crystallizing in the monoclinic system, the latter in the triclinic system; and the members of the solid-solution series of the plagioclase feldspars. The member of the plagioclase series which contains in excess of 90 per cent sodium alumino-silicate ($NaAlSi_3O_8$) is known as albite; feldspars rich in albite are called soda feldspars. The plagioclase feldspar which contains in excess of 90 per cent calcium alumino-silicate ($CaAl_2Si_2O_8$) is called anorthite. Feldspars rich in anorthite are calcic feldspars. The potash and soda feldspars are collec-

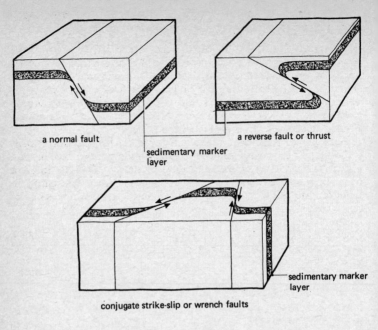

a normal fault

sedimentary marker layer

a reverse fault or thrust

conjugate strike-slip or wrench faults

sedimentary marker layer

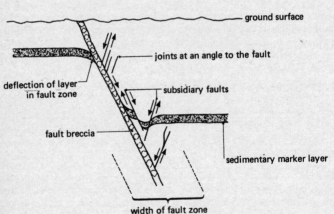

ground surface

joints at an angle to the fault

deflection of layer in fault zone

subsidiary faults

fault breccia

sedimentary marker layer

width of fault zone

Figure 26. Different kinds of geological faults.

tively known as the alkali feldspars; they are relatively rich in silica (SiO_2) and are characteristic minerals in acid igneous rocks. By contrast the calcic feldspars, which contain a smaller proportion of silica, are abundant in basic igneous rocks. Although the feldspars are important minerals in most igneous rocks, they are less abundant in most sedimentary rocks except in the variety of sandstone known as arkose. Feldspar also occurs in some metamorphic rocks.

Feldspars of commercial value are mainly alkali feldspars, principally orthoclase. They occur in sufficient concentration to be worked in some PEGMATITES. Some of the individual crystals of feldspar in some pegmatites are a few tens of metres in diameter. Calcic feldspar is the dominant mineral in the igneous rock ANORTHOSITE, which is related to gabbro. There is some interest in working anorthosites for their feldspar content.

Feldspar is used in the manufacture of GLASS, ceramic glazes, PORCELAIN, enamels and artificial teeth. Some feldspar is also used as an ABRASIVE. Production in 1971, in tonnes, was: USA, 673868; West Germany, 353693; USSR, 249000. (♢GRANULES; MINERAL FILLERS.)

Feldspathoid minerals. Although members of this pale-coloured group of minerals are related to the FELDSPARS, they differ from them in structure and in being deficient in silica (SiO_2). Feldspathoid minerals commonly occur in intermediate igneous rocks which are low in silica or abnormally rich in alkalis. Two feldspathoid minerals of economic importance are LEUCITE and NEPHELINE.

Felsic minerals. ♢IGNEOUS ROCKS.

Felsite. The name felsite, although somewhat archaic, remains a useful field term to describe many fine-grained or cryptocrystalline, pale-coloured igneous rocks of acid composition. Some are devitrified volcanic glasses. Quartz (♢SILICA) and alkali FELDSPAR are the principal minerals in felsites. Many felsites yield excellent CRUSHED STONE for road making (♢ROAD METAL).

Ferrous minerals. A term used loosely by economic geologists to describe not only the minerals of IRON (Fe) but also those of the metalliferous elements which are alloyed with iron to produce a large variety of iron and steel products. The group includes chromium, cobalt, columbium, manganese, molybdenum, nickel, rhenium, silicon, tantalum, tungsten and vanadium minerals.

Fertilizer sources. Plants require nitrates, phosphates and potash for life. The mineral sources of these essential ingredients are supplied by NITRATE DEPOSITS, PHOSPHATE DEPOSITS and POTASH DEPOSITS. In addition to these primary nutrients, secondary nutrients are also required. They include calcium, magnesium, silicon, chlorine and sulphur. Calcium is normally added by spreading granulated limestone or lime. Sulphur is usually added in the form of sulphates. Suitable magnesium compounds are supplied after processing dolomite, magnesite, olivine or the various MAGNESIUM SALTS. Boron is another minor, but essential nutrient, supplied from processed BORATE DEPOSITS. Plants also require traces of certain metals such as copper, iron, cobalt, zinc, manganese, vanadium and molybdenum. There is much current research aimed at understanding the distribution and value of these trace metal requirements of good soils. Soil may be conditioned by the addition of sand and peat.

Field capacity. The amount of water held in the soil by capillarity after excess GRAVITATIONAL WATER has drained away and after the rate of downward movement of water has materially decreased. Field capacity is thus synonymous with SPECIFIC RETENTION in the SOIL WATER ZONE.

Filter sand. Washed and graded SAND or GRAVEL which can be used as a filtration medium, as for example in the treatment of sewage and water. Suitable sands for rapid filtration consist of coarse-grained, well-rounded quartz grains (◊SILICA) and should contain a minimum of silt, clay, organic material, flat rock fragments, and iron or manganese oxides. Finer-grained sands are used for slower filtering; they need not consist of rounded grains, but the other requirements should be met.

Good filter sands are uncommon and deposits may require lengthy and expensive processing before use.

Fines. ◊GRAVEL.

Fireclay. Fireclays, characteristically developed beneath coal seams, are deficient in alkalis, lime, magnesia and iron oxides, and thus form suitably REFRACTORY materials. The principal CLAY MINERAL in many fireclays is kaolin. One theory explaining the development of fireclays is that they are seat-earths, or fossil soils in which the plants of a swamp grew. Fireclay is characteristically unbedded and may show rootlet remains. If the soil in which the swamp plants grew was essentially sandy, the resultant seat-earth rock is commonly known as a ganister (a variety of sandstone); whereas if the soil was clayey, the seat-earth is called a fireclay or underclay. Fireclays and ganisters are deficient in alkalis possibly because of leaching by plants.

Fireclays commonly occur beneath coal seams in many coal-bearing rock successions as for example in the Upper Carboniferous or 'coal measure' sequences of North America and Europe. Both in the USA and Britain fireclays in the coal measures have been worked for their refractory value.

Most fireclays are highly plastic. Non-plastic fireclays are known as flint clays because of their dense texture and distinctive fracture. Flint clays are generally mixed with plastic clays before use. Some flint clays contain diaspore which enhances their refractory properties. Good-quality fireclays are used for making refractory products; less high-grade fireclays are used in the manufacture of sanitary earthenware. (◊BONDING CLAYS; FOUNDRY SAND; POTTERY AND STONEWARE CLAYS.)

Fissile. ◊MUD AND MUDSTONE.

Fissure. ◊JOINT.

Flagstones. ◊SANDSTONE.

Flint clays. ◊FIRECLAY.

Floods. Situations when a river is at an unusually high stage, the stage at which the stream channel becomes filled and where the flow overtops the banks. Inasmuch as the banks of a stream vary in height throughout its length, there is no definite stage at which a river can be said to be in flood. As floods result primarily from surface RUNOFF, permeable catchments are rarely subjected to flooding. In other catchments flooding may follow a period of intense rainfall or the melting of accumulated snow or a combination of melting snow and rainfall.

Flow duration curves. A curve constructed from cumulative amounts of river flow versus corresponding units of time. Thus the curve will show the percentage of time that any discharge was equalled or exceeded, and the shape of the curve provides a clear picture of the nature of the river flow. A flat curve results from a river subjected to few floods and a large BASE FLOW component whereas a steep curve indicates a flashy river, subject to very low flows.

Fluoride. A minor chemical constituent of ground water, which is important because it affects teeth. Concentrations in excess of 4 ppm can produce dental fluorosis, yet too little fluoride may be detrimental in that it can produce mottled teeth in children. Fluoride values of about 1 ppm are known to inhibit dental caries and in some water supplies are artificially maintained at this level.

Few water samples have been analysed that yield more than 10 ppm fluoride. The natural concentration of fluoride appears to be limited by the solubility of FLUORITE (CaF_2), which is about 9 ppm fluoride in pure water. High, natural fluoride levels in chalk and Jurassic LIMESTONE AQUIFERS in England are associated with soft alkaline ground waters. Waters with high calcium concentrations are noted for their low fluoride content of not more than 1 ppm.

Fluorite. Calcium fluoride (CaF_2) commonly occurs as translucent cubic crystals which when they contain impurities may be yellow, green, pink, purple, blue or brown. The mineral fluorite is often known commercially as fluorspar. A yellow- and blue-banded variety, suitable for ornamental use, is called Blue John.

Fluorite is abundant in many hydrothermal veins from which it may be worked as the principal economic mineral, or it may be one of the GANGUE minerals in a vein containing metalliferous ores. Veins containing fluorite cut the Carboniferous limestone where it crops out in Derbyshire, Yorkshire and Durham. Major worked deposits of fluorspar replacing calcite occur in veins and beds in Mississippian limestones in southern Illinois and western Kentucky. Some residual accumulations of fluorite, derived from vein or replacement masses, have also been exploited.

Most fluorite is employed as a flux in iron and steel making. Among its many other uses are the manufacture of hydrofluoric acid and its derivatives, its use in the ceramic industry, its employment for the separation of URANIUM-235 from uranium-238, its use in the optical glass industries, and its use in the manufacture of organic FLUORIDES, for example those employed in aerosol sprays. Production in 1974, in tonnes, was: Mexico, 1 180 000; USSR, 450 000; USA, 225 000.

Fluorspar. ◊FLUORITE.

Fold. A curve, bend or deflection imposed on a set of pre-existing surfaces or layers in a rock mass. The process of developing a fold is called folding. Although most of the folds recognized by geologists are deformed sedimentary layers or beds (◊SEDIMENTARY ROCKS), the name fold is equally applicable to similar deflections produced in other surfaces such as CLEAVAGE. The majority of folds are the result of the lateral compression of originally horizontal rock layers at a time when they were capable of flowing. Some folding is also caused by the gravitational gliding of large rock masses, and by density instabilities within a rock sequence. Curved linear belts of the earth in which there are many folds and other indications of intense deformation are known as orogenic zones. The most recent orogenic zones are the Circum-Pacific and Alpine-Himalayan fold mountain belts developed during the Mesozoic and Cenozoic eras. Orogenic belts evolve in the contact zones between converging lithospheric plates (◊THE EARTH). Many geologically ancient orogenic zones are no longer areas of high mountains, denudation over millions of years having lowered them.

When the rocks in the core of a fold are older than those of its envelope, that fold is an anticline, and when they are younger the fold is called a syncline. Most anticlines are upfolds and most synclines are downfolds. The limbs of a fold are separated by areas of maximum curvature known as hinge zones. When the hinge line of a fold is inclined to the horizontal the fold is said to plunge. The imaginary plane which bisects the angle between the limbs of a fold is called an axial plane. The axial planes of folds range in dip from horizontal to vertical. A simple scheme for classifying folds according to the dips of their axial planes is shown in Figure 27. Folds vary in size from those which can be seen only under the microscope to those whose dimensions from one hinge line to another are tens of kilometres, and which

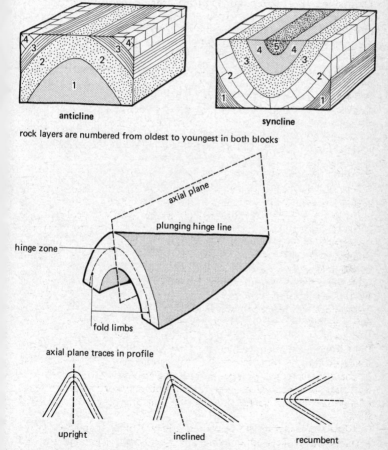

anticline

syncline

rock layers are numbered from oldest to youngest in both blocks

axial plane

plunging hinge line

hinge zone

fold limbs

axial plane traces in profile

upright   inclined   recumbent

Figure 27. Classification of folds according to the inclination of their axial planes. Folds represented in profile by a single layer.

may extend along their hinges for hundreds of kilometres. After denudation the presence of a fold may not be reflected in the topography of the area which it underlies. Some rocks have been folded more than once, and as a consequence a complex pattern of OUT-CROPS results.

The size, shape and distribution of folds largely controls the distribution of rocks of different age, hence the distribution and depths of many valuable mineral deposits.

Foliation. ▷CLEAVAGE IN ROCKS.

Formation. The fundamental unit in rock classification: a body of rock generally of homogeneous character which may be traced across country in which it differs from formations above and below it.

Forsterite. ▷OLIVINE.

Fossil fuels. The preservation of once-living matter in sediments which in their turn become SEDIMENTARY ROCK gives rise to the fossil fuels, of which the most common are COALS, oil and gas (▷PETROLEUM). Coal is derived from the compression and alteration of masses of plant debris. Oil and gas result from the transformation of animal and plant matter which was originally present in sea-floor sediment and subsequently affected by pressure and heat.

It has been said that by AD 2000 most of the world's fossil fuels will have been consumed, and that 80 per cent of them will have been burnt in the last 300 years. The doubling period for the consumption of fossil fuels is about ten years. Estimates of reserves differ very widely with a minimum estimate of the lifetime of reserves being about 10 years and the major being somewhat over 50 if the rate of consumption does not change. Meadows *et al.* have suggested the effects of increases in world fuel reserves and concluded that present (1972) rates of consumption could not long continue. Changes in fuel consumption and some CONSERVATION measures have been introduced for political as well as economic reasons as a result of action by the Organization of Petroleum Exporting Countries (OPEC), but the long-range effects of these are yet to be calculated. They have prompted intensified exploration for fossil fuels in parts of the world other than the Middle East

The calculated effects of increases in global fuel reserves

Resource	Known reserves	Lifetime at present consumption (yr)	Estimated increase in consumption (%/yr)			A: lifetime at average estimated rate of increase	B: lifetime as in A but assuming 5 times the known reserves
			Mini-mum	Aver-age	Maxi-mum		
Coal	5×10^{12} tons	2300	3·0	4·1	5·3	111	150
Natural gas	$1·1 \times 10^{15}$ ft³	38	3·9	4·7	5·5	22	49
Oil	455×10^9 bl	31	2·9	3·9	4·9	20	50

(After Menard.)

and increased attention being given to other sources of ENERGY. (Figure 28.)

D. H. Meadows, D. L. Meadows, J. Randers and W. Behrens, *The Limits to Growth*, Pan, 1972.

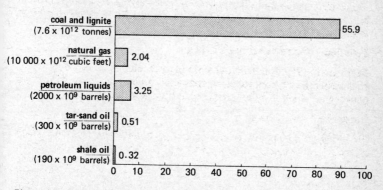

Figure 28. Fossil fuels, their energy content in percentages of the world total recoverable supply and in units of 10^{15} k Wh_{th}. (After Menard.)

Foundry sand. The REFRACTORY properties of silica SAND are utilized in several foundry processes. Sand which is suitable for contact with molten metal should be rich in anhydrous SILICA but free of opaline silica, relatively fine-grained, highly permeable to allow the escape of gases, capable of resisting sintering, strong and durable. Core sand should be washed, graded, of high permeability and contain little clay. Furnace bottom sand used for lining furnaces should contain sufficient clay bond to provide a cohesive material, and should contain a range of particle sizes from clay to granules. Naturally bonded sand contains sufficient indigenous clay to allow moulds to be made. Suitable unconsolidated sand–clay mixtures are worked in some eastern States of America. Processed moulding sand is washed and graded silica sand which can be mixed with a bonding clay.

Many sands and SANDSTONES can be exploited successfully for foundry sands. British examples include some ganisters of Carboniferous age, the Folkestone Beds of Cretaceous age, and the Thanet sands of Tertiary age. Other refractory minerals such as olivine may also be used in foundry sands.

Fracture. A fracture is a surface in a rock mass which has resulted from failure. When a brittle rock is subjected to a critical level of stress it fails by forming either a single set of extension fractures or a conjugate system of shear fractures (Figure 29). The name fracture covers a broad spectrum of types of discontinuities including JOINTS, FAULTS, VEINS and fissures.

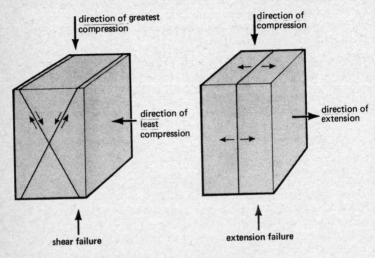

Figure 29. Fracture in rocks.

Fracture cleavage. ⬦CLEAVAGE IN ROCKS.

Franklinite. A complex iron, zinc and manganese (Fe, Zn and Mn) oxide of the SPINEL group. The mineral occurs associated with zincite and willemite, in the commercially important zinc deposits of New Jersey, USA.

Freshwater. Water containing between 0 and 1000 ppm of total dissolved solids. Freshwater barriers comprise freshwater injected to stop inflow of saline water into COASTAL AQUIFERS.

Fuel. Combustible materials to provide heat and light. They have been sought by man since his earliest days. Today they fall into three major categories: FOSSIL FUELS, nuclear fuels and renewable fuels.

G. Holister and A. Porteous, *The Environment: A Dictionary of the World Around Us*, Arrow Books, 1976.

Fossil fuels include COAL, oil and gas (⬦PETROLEUM); nuclear fuels are those such as certain forms of URANIUM which are radioactive and generate heat by atomic decay. Renewable fuels are those largely produced by the plant kingdom, primarily wood of various kinds, but most organic substances may be treated and utilized in energy-producing processes. The generation of the gas methane from farmyard manure is one such process which has been brought into play where or when fuel is scarce.

Fossil fuels are the most important source of power today and will continue to be so till the end of this century but thereafter their supply will so diminish that other forms and sources will become necessary. (◊CONSERVATION; ENERGY.)

Fuller's Earth. ◊ADSORBENT AND BLEACHING CLAYS; BENTONITE.

G

Gabbro. A dark-coloured, coarse-grained, plutonic igneous rock of basic composition, commonly containing calcic-plagioclase feldspar, augite or another PYROXENE; some may contain olivine. In composition they are the equivalent of basalt. Most gabbro intrusions are small. Gabbros are mainly worked for CRUSHED STONE although some have been used as DIMENSION STONE.

Galena. Lead sulphide (PbS) containing 87 per cent lead. It has a metallic lustre, is lead grey in colour, and commonly occurs as cubic crystals showing perfect cleavage. The mineral is soft (hardness 2·5) and has a very high SPECIFIC GRAVITY (7·5). It is commonly found in veins and as replacement deposits in limestones. Galena is the most important ore of lead and often contains sufficient silver to become a silver ore.

Gallium (Ga). An element more abundant (the continental crust contains 17 ppm) than antimony, silver, molybdenum and tungsten, but for which there is little demand. It is used in tiny amounts by the electronics industry and is derived solely as a by-product of certain aluminium and zinc ores. World production (1973) is speculative, but probably in the order of 15 tonnes.

Gamma logs. Gamma ray logging is based on measuring the natural radiation of gamma rays from certain radioactive elements that occur in varying amounts in geological formations. Present-day continuous gamma ray logging of wells, both water and oil, using an instrument known as a scintillation counter, is almost a routine operation following the drilling of the well. The log is a diagram showing the relative emission of gamma rays, measured in counts per second, plotted against depth below the surface. In appearance the resultant curve is not unlike that of an electric log. (◊ELECTRICAL LOGGING.)

Some rocks, such as clay, contain more radioactive elements, such as uranium, thorium and the radioactive isotope of potassium (^{40}K), than others such as sand, sandstone and limestone. Thus a gamma log, particularly of unconsolidated materials, will delineate clay (relatively impermeable) from the more permeable materials and the distinction is clearer than in an electrical logging method.

The gamma ray log, therefore, like the electrical log is especially valuable in FORMATION correlation but its great advantage over electric logging is that it can be used in cased, as well as uncased, wells.

Gangue. The material associated with an ore mineral which is regarded, at the time of working, as waste material (pronounced 'gang'). At a later date, discarded spoil-heaps of gangue may become of economic value.

Ganister. ⋄FIRECLAY; SANDSTONE.

Garnet. A group of minerals crystallizing in the cubic system. Only two are of economic significance: the most important is almandine garnet ($3FeO.Al_2O_3.3SiO_2$); andradite ($3CaO.Fe_2O_3.3SiO_2$) is of lesser significance. Garnets are relatively hard (6·5 on Mohs' scale), and are thus mainly used as ABRASIVES.

Although garnets are common in many medium-grade regional and contact metamorphic rocks they are also present in minor quantities in many igneous and sedimentary rocks. They are concentrated in economic quantities in only a few locations. The best known is that in the Adirondack Mountains of New York State where large crystals in a rock containing up to 70 per cent almandine garnet are worked. On crushing, these crystals break into fragments with especially sharp edges. In some so-called 'black sands' garnet is concentrated as a PLACER DEPOSIT.

Garnet is used in garnet paper, for polishing hardwood and paint, and in abrasive wheels used for cutting slate or marble. The hardness of garnet can be improved by heat treatment. Some garnet is of gem quality. Production in 1971, in tonnes, was: USA, 17222; India, 1557.

Garnierite. ⋄NICKEL.

Gemstone. A mineral or mineral fragment which is used for personal adornment. Desirable properties in a gemstone are a combination of beauty (subjective but usually based on colour, transparency and lustre), durability and rarity. Diamond, ruby and sapphire (varieties of corundum) and emerald (a variety of beryl) are usually termed precious; all other gemstones are termed semiprecious. Agate, amethyst, cairngorm, citrine, crystal, jaspar, onyx and rose are all varieties of quartz.

Gemstones, especially the precious stones, are usually artificially cut and polished. Many gemstones can be made synthetically.

Geological cycle. ⋄THE EARTH.

Geochemical exploration. The search for mineral deposits or petroleum by detection of abnormal concentrations of elements or hydrocarbons in surficial materials or organisms, usually by instrumental, spot-test or rapid field techniques.

A. A. Levinson, *Introduction to Geochemical Prospecting*, Applied Publishers Ltd, Calgary, 1974.

Geochemical prospecting. The search for concealed deposits of ore by the systematic analysis of soils, stream sediment, stream water or vegetation for abnormal concentrations of metal. Modern chemical reagents often permit semi-quantitative analyses of trace amounts of metal to be done in the field. The results of the analyses are plotted on a map and areas of abnormal concentration ('anomalies') delineated from the normal 'background' value. Anomalous areas may be followed up by drilling.

Geophysical exploration, Methods. Geophysical exploration affords a means, at moderate cost in relation to potential returns, of either directly locating subsurface resources or of determining the subsurface structure so that exploratory drilling may be restricted to favourable areas and wildcat drilling can be avoided. Rapid development of prospecting geophysics over the last few decades has been largely due to the ever-expanding need for oil, and refinements in geophysical techniques have been geared to the needs of the oil industry.

Geophysical exploration comprises an assortment of methods utilizing physical principles to enable the prospector to 'see' underground; these methods are based on the fact that rocks, minerals and fluids of different types may display wide variation in certain physical properties such as elasticity, magnetism, density, electrical characteristics and natural radioactivity levels. These physical properties form the basis of the five principal methods of geophysical exploration: seismic reflection and refraction (◇SEISMIC PROSPECTING), gravity, magnetic, electrical and radiometric. Some of the methods are primarily employed at, or above, the land surface and occasionally in underground workings while others have been mainly adapted for subsurface (downhole) work. (Figure 29.)

Of the surface methods, seismic reflection is the most widely used in oil and gas exploration with gravity, seismic refraction and magnetic methods as most useful ancillary procedures. In the search for ore minerals the magnetic, electrical and radiometric methods are most commonly used. In ground water exploration the surface techniques almost exclusively employed are electrical and shallow seismic refraction, but in the search for this resource subsurface techniques are much more important.

In subsurface (downhole) geophysical work the measurements are made by lowering specially encased miniaturized instruments (SONDE) into wells. The records produced provide an invaluable supplementary tool to the geological log of the well. Seismic, magnetic and gravimetric measurements can all be made downhole but paramount in importance in the location of fluids in subsurface RESERVOIR ROCKS are the electrical and radiometric methods. These downhole geophysical methods do not directly indicate the type, porosity or permeability of

96

the rocks but they do measure other properties that vary with the factors that determine whether or not a formation may be sufficiently porous and permeable to serve as an aquifer or reservoir rock.

M. B. Dobrin, *Introduction to Geophysical Prospecting*, McGraw-Hill, 1960.

Geothermal energy. Heat from within the crust of the earth, usually expressing itself in emissions of hot water (GEYSERS), steam and gas. It is commonly associated with volcanic regions of the world, but warm springs occur at many places far removed from such regions. High temperatures are also found in many deep mines.

The use of geothermal or volcanic heat for domestic or industrial purposes is relatively recent, having begun in Italy in 1904. Among the seventeen other countries now actively encouraging the use of this source of power are New Zealand, Japan, the USA, Iceland and the USSR. The United Nations Organization is sponsoring exploration projects in El Salvador, Chile and Turkey. (Figure 30.)

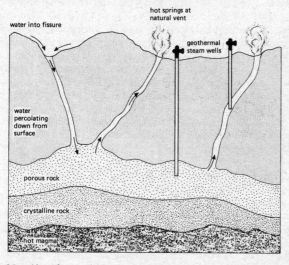

Figure 30. Heat from within the crust raises the temperature of water percolating down from the surface. Steam may reach the surface through natural vents to give rise to hot springs.

In all the 'hot areas' shallow (volcanic) igneous rocks are thought to occur within 8 kilometres of the surface and to be perhaps still in a molten or partly molten state, and major rock FRACTURES are an essential structural element. The issuing hot water or steam is principally rainwater which has penetrated down to the hot regions from the surface. It may be used to generate electricity or to give heat. Most electricity-generating plants so far are rather small (10–13 milliwatts). One of the problems attendant upon the spent fluid is that it commonly contains pollutants such as boron, chlorine, fluorine or arsenic which cannot be released into surface drainage.

Impetus to the developing of geothermal power plants has been given by the rising costs of FOSSIL FUELS. ($\diamondsuit$ENERGY.)

Germanium (Ge). Used principally as a semiconductor in the electronics industry. Germanium is recovered commercially as a by-product of zinc ores.

Geyser. An intermittent thermal spring resulting from the expansive force of superheated steam produced by water in contact with hot rock at depth in the earth's crust. ($\diamondsuit$GEOTHERMAL ENERGY.)

Glass. Common window glass is a supercooled silicate liquid comprising about 70 per cent silica (SiO_2), 14 per cent lime and magnesia ($CaO + MgO$), 12 per cent soda (Na_2O) and 1–2 per cent alumina, ferric oxide and sulphur trioxide. Special varieties of glass contain other compounds. For example, heat-resistant or 'pyrex' glass contains borates, which greatly reduce the coefficient of expansion; flint glass of high brilliance, used for making cut glass, contains lead oxide (PbO_2); and quartz glass, of use in laboratories, is silica-rich. In optical glass the potash (K_2O) content is twice that of the soda content. About 7 or 8 per cent of potash plus lead oxide is added to make crystal glass. In small quantities the presence of certain elements colours glass. For example, gold is used to give a ruby colour, cadmium a yellow colour, and manganese a violet colour. Brown glass, often used for bottles, is coloured by iron impurities. The addition of barium minerals increases the fluidity of molten glass, and thus facilitates the manufacture of intricate glassware.

As the above outline suggests, the principal raw material used in glass making is quartz SAND ($\diamondsuit$SILICA). The specifications a sand deposit has to fulfil to be suitable for glass making are generally more rigorous than for most uses of sand. The silica content should be 95–99·8 per cent, that is the sand should be almost pure quartz. The sand grains should be well sorted. In the main, unconsolidated sands or very weakly cemented sands are best.

An especially pure sand of Upper Cretaceous age at Loch Aline in the Morven area of the Scottish Highlands is an example of an excellent British glass sand. Other glass sands in Britain are also of Cretaceous age; they include the Folkestone Beds, Tunbridge Wells sand and Ashdown sand of the Wealden area in south-east England.

In the USA two important sources of glass sand are obtained from the Oriskany quartzite of Devonian age in some Appalachian States, and from the St Peter sandstone of Ordovician age in Illinois, Missouri and Arkansas.

Glass silk. ⟡MINERAL AND ROCK WOOL.

Glauber's salt. ⟡SODIUM CARBONATE AND SULPHATE.

Glauconite. A complex mineral (approx. $K(FeAl)_2(SiAl)_4O_{10}(OH_2)$), characteristic of some sandstones deposited in the sea. It has few if any uses, but it is a potential source of POTASH and could possibly be used for ion exchange. Unweathered glauconitic SANDSTONES are generally green, and weather brown or orange. Glauconitic sandstones are abundant in the Cretaceous succession of south-east England, and beneath the New Jersey coastal plain.

Gneiss. A coarse-grained, banded, crystalline rock. The mineralogical composition is commonly like that of GRANITE, i.e. composed of feldspars, micas and quartz. Strong pressures and high temperatures are required for the formation of gneisses which are thus metamorphic rocks produced by the alteration of previously existing rocks. Common in the western highlands of Scotland and the Hebrides, in the Canadian Maritime Provinces and the Canadian Shield, in Scandinavia and the Alps, gneiss is little used economically except locally for ROAD METAL and ballast. (⟡CRUSHED STONE; DIMENSION STONE.)

Goethite. ⟡GOSSAN; IRON; IRONSTONES.

Gold (Au). A scarce, precious metal which occurs in nature mainly in the native state. The soft, but virtually corrosion-resistant, yellow metal is almost indestructible. Gold has been mined for over 6000 years and was one of the earliest metals used by man. It provides an almost perfect example of man's ability to recycle a product – if he wishes to. Much of the gold that has ever been mined is still in circulation.

Gold has a long history as a medium of exchange. First used as jewellery, then in coinage, it was finally stockpiled by governments to validate paper money. For decades the price of gold was maintained by the US government at $35 an ounce; recently its official price has been abolished and a two-tier price system operates. There is a free market price for industrial gold (this attained $186.50 in 1974), and an official price ($42.22 an ounce) agreed by members of the International Monetary Fund.

For use in jewellery and ornaments, gold is usually alloyed with BASE METAL. The term carat means a 24th part and is used to express the proportion of gold in the alloy. Thus an 18-carat gold ring contains $\frac{18}{24}$ or 75 per cent gold.

Gold occurs in a variety of geological environments but the most important are veins or LODES and, because of its indestructability, in

PLACER DEPOSITS. Gold is also recovered as a by-product or co-product of ores mined essentially for copper and other base metals.

Present-day gold production is dominated by South Africa which produces about two-thirds of world production. The gold comes from the very ancient (2200 million years) Witwatersrand ('the Rand') sedimentary basin – a probable fossil placer deposit. Other major producers include the USSR, Canada and the USA. Total world production (non-Communist) in 1974 was about 1000 million tonnes.

Gossan. The leached and oxidized surface or near-surface part of a sulphide vein. It usually consists of a cellular deposit of hydrated iron oxide (goethite). 'Iron hat' is a synonym. Some experienced prospectors maintain that the structure and colour of the gossan is an indication of the nature of the underlying sulphides. The gossan occasionally contains a concentration of highly insoluble elements such as gold.

Gouge. ◊FAULT.

Grade. The quality of an ORE (1); the metal content of an ore. Terms such as 'high' grade and 'low' grade are subjective but tend to reflect average economic metal contents of an ore at any particular time. (◊CUT-OFF.)

Granite. In the stone trade the name granite has been used to describe a wide range of hard and durable rocks, not all of which are granite according to even the broadest geological definition. To the geologist, granites are coarse-grained, plutonic igneous rocks of acid composition. The majority occur in major igneous intrusions known as BATHO-LITHS. All rocks belonging to the granite clan contain quartz (◊SILICA), FELDSPAR and MICA. According to the percentage of the type of feldspar, the rocks of the granite clan are divided into: alkali granites, those containing greater than two-thirds of alkali feldspars; adamellites, those containing between one-third and two-thirds of alkali feldspars; and granodiorites, those containing greater than two-thirds of calcic-plagioclase feldspars. With increasing calcic-plagioclase content granites become less 'acidic'.

Most granites are pale-coloured rocks, and some contain attractive large crystals of feldspar, thus making them suitable ornamental or facing stones. When granite is quarried for DIMENSION STONE it is commonly extracted in blocks bounded by JOINT planes. The blocks may then be shaped by splitting the stone along certain planes of weakness which are not easily detected except by experienced workers. The easiest direction of splitting is called the rift, and the second easiest direction, at right angles to the first, is known as the grain or run. Along the direction at right angles to the rift and grain, splitting is difficult; this direction is called the hardway or head grain. The rift and grain directions follow planes of micro-fractures, or planes along which minute bubbles in the crystals are concentrated. Granite which is unsuitable for dimension stone is often worked for CRUSHED STONE or RIP-RAP. (◇ABRASIVES; GRANULES; ROAD METAL.)

Granites are abundant in many areas of the world underlain by Precambrian rocks and in many ancient or present-day fold mountain belts. Their large bulk and relatively low monetary or unit value mean that they rarely enter international trade statistics. Well-known granites in Britain which are worked for dimension or crushed stone include the Dartmoor and Sharp granites in England, and the Rubislaw and Peterhead granites in Scotland.

Many granites are also of economic interest because of their association with ore deposits and CHINA CLAY.

Granules. A term used to describe granulated material spread on the surface of ASPHALT products which are employed for roofing and similar purposes. Felt is saturated with asphalt and its surface is coated with crushed minerals or rocks. The granulated material, especially TALC, prevents adhesion while the felts are in rolls, improves fire and weather proofing, and increases the attractiveness of the roof. Houses roofed in material of this type are common in the USA.

A wide variety of strong, durable, non-porous rocks and minerals are used as granules, including granite, rhyolite, dolerite, basalt, sand, sandstone, gravel, dolomite, slate, marble, feldspar, mica and talc. In addition shells and artificial products such as crushed brick and slag have also been used. To increase their attractiveness some granules are artificially coloured.

Graphite. Native CARBON (C) in a hexagonal crystalline form, very soft (hardness 1–2), of metallic lustre, black, flaky and opaque. It was originally called plumbago, being mistaken for lead.

Graphite occurs as disseminated flakes in some schists, gneisses and marbles. Where it is sufficiently concentrated in these rocks, it can be worked for flake graphite as in the USA, Madagascar, Norway, West Germany and Austria. Commercial quantities of graphite also occur in some veins and pegmatites, the deposits in Ceylon being the best known. In Mexico important graphite deposits have resulted from the thermal METAMORPHISM of coal seams cut by granite DYKES. Metamorphosed coals are also worked in Italy and Austria.

Graphite is used for making crucibles for metallurgical purposes, as a lubricant, in paints and pigments, in the manufacture of rubber, pencils and batteries, as a moderator in atomic reactors, and as brushes in electrical equipment. Production in 1971, in tonnes, was: South Korea, 375172; Mexico, 50916; China, 30000. Synthetic graphite is manufactured from petroleum coke by electrical furnace processes.

Gravel. A loose, unconsolidated sediment containing particles (often pebbles or cobbles) in excess of 2 millimetres in diameter. Most gravels also contain sand and fines, that is material of clay and silt size ($\diamond$ MUD AND MUDSTONE). The name shingle is used to describe a gravel without an appreciable proportion of sand or fines. The commonest rocks found as pebbles or cobbles in gravels are granite, quartzite, sandstone, limestone and flint. They are all strong, even-textured rocks.

The principal use of gravel is for AGGREGATE. The pebbles are generally well rounded and mixed with sharp sand, that is sand containing angular grains of quartz. For many practical purposes pebbles greater than 4 centimetres in diameter are unwanted. ($\diamond$ FILTER SAND; GRANULES.)

In Britain the majority of worked gravels are superficial deposits of Pleistocene or Recent age. The principal environments in which they have been laid down are those of the middle courses of large rivers, and former glacial outwash regions. Some river gravels are now isolated in terraces at higher levels than the present-day channel of a river as a consequence of falling sea levels since the Tertiary. Fluvio-glacial gravels which are now at a shallow depth beneath the sea are being investigated in some British offshore areas. In general, submarine sands and muds of Pleistocene age are more abundant. Important Quaternary gravel deposits are worked in the Thames and Trent valleys.

British bedrock gravels are excavated from the Bunter pebble beds of Triassic age in the Midlands of England. These former continental deposits are only weakly cemented, and thus the pebbles and sand are easily separated.

Gravels are always worked in opencast pits by diggers or draglines. After extraction they require washing and screening. Some old gravel pits are used as water-sport centres or as nature reserves.

Gravel pack. A gravel screen or envelope surrounding perforated CASING in a well. The reasons for emplacing a gravel pack in a well include prevention of inflow of fine-grained sediment into the well during pumping and protection of the casing from caving of surrounding aquifer materials. The gravel pack increases the effective well diameter and in unconsolidated aquifers the well containing the pack will have a higher SPECIFIC CAPACITY than unpacked wells.

Gravimeter. A geophysical instrument to measure gravity, especially gravity anomalies. Gravimeters record the direct effects of the pull of gravity on a weight suspended by a delicate spring. Variations in the length of the spring relate directly to the vertical intensity of the gravity field. Accuracies of about 10^{-8} of the field of gravity can be made using this instrument. Under ideal circumstances, after many corrections have been made, quantitative results will support the possible geological nature or structure of the local vicinity.

Gravitational water. Excess soil water which drains through the

SOIL WATER ZONE, and indeed through the AERATION ZONE, under the influence of gravity (Figure 7). Thus, this water moves downward under its own weight, towards the WATER TABLE.

Greenockite. ⟡CADMIUM.

Greywackes. ⟡SANDSTONE.

Grit and gritstone. Names commonly, but loosely, used to describe some coarse-grained SANDSTONES, especially those containing angular grains of quartz, and rough to the touch. Some well-cemented grits are used as DIMENSION STONE, and many more can be worked for CRUSHED STONE. Weakly cemented grits can often be used as a source of sharp sand (⟡AGGREGATE). In Britain the name millstone grit is used to describe a sequence of coarse sandstones and inter-bedded mudstones (⟡MUD AND MUDSTONE) of Namurian (Upper Carboniferous) age in the Pennine Hills and elsewhere. As the name suggests these grits were used for making millstones or grindstones. Some coarse, rubbly, fossiliferous LIMESTONES have been called grits.

Ground water (phreatic water). Water occurring within the ZONE OF SATURATION of an aquifer is the only part of all subsurface water which is properly referred to as ground water or phreatic water. Ground water present within this zone can be considered to be occupying a large natural reservoir or series of reservoirs whose capacity is the total volume of the PORES or VOIDS in the rocks that are filled with water.

Ground water may be of variable chemical quality ranging from wholesome potable waters to highly mineralized BRINES. It can be correctly considered as a mineral resource, and its uniqueness is due to the fact that it is the only replaceable resource.

Extensive aquifers comprise ground water basins, and the direction of flow of the ground water and delineation of the discharge points may be evaluated by studying the ground water levels in wells sunk into the zone of saturation. Commonly resource evaluation is undertaken on ground water catchments which may or may not be of similar dimensions to the overlying surface water catchments in UNCONFINED AQUIFERS. Ground water contours may be drawn as lines of equal water level throughout an aquifer, and ground water flow directions are always at right angles to the contours.

Ground water development can be undertaken via wells or works at a spring head, and various forms of models can be constructed to simulate ground water conditions within an aquifer and to indicate optimum development procedure.

Grout. Grouting or cementing well CASING involves filling the space around the pipe, usually between the pipe and the well wall, with a suitable slurry of cement or clay. Cement grout is a mixture of Portland cement and water. Puddled clay may also serve as a grout provided it is used at a depth where drying and shrinkage of the clay will not occur. Grouting the upper part of a water well is most important in the protection of that well from pollution.

Guano. ⟡NITRATE DEPOSITS; PHOSPHATE DEPOSITS.

Gusher. The eruptive discharge of oil, gas or water from a drilled well by the pressure exerted within the RESERVOIR ROCK. Oil wells are ideally engineered to allow the oil to emerge under its own pressure, but unexpectedly high pressures may destroy the drilling rig if encountered without proper precautions.

Gypsum. A monoclinic mineral ($CaSO_4 2H_2O$) which is white or colourless. Rock gypsum is the impure, commonly red-stained granular form found in layers; gypsite is the impure earthy form which in the USA occurs in association with marls in surface deposits; alabaster is the pure, fine-grained, commonly translucent form which can be worked ornamentally; satin-spar is the fibrous silky form characteristically found in veins; and selenite is the form which occurs as transparent crystals in some clays and mudstones.

The majority of gypsum of economic significance occurs as beds in association with other minerals characteristic of EVAPORITE DEPOSITS derived from bodies of seawater. If the temperature of the water from which crystallization is occurring exceeds about 25°C, ANHYDRITE rather than gypsum is precipitated. Gypsum may, however, result from the later hydration of anhydrite. It may also be formed during the conversion of limestone to dolomite. In some arid areas where gypsiferous strata are being weathered there are wind-blown dunes of gypsum. In more humid areas where gypsum is being dissolved it may then be recrystallized by efflorescence when ground waters migrate towards the surface, perhaps during hot, dry summers. This process gives rise to gypsite deposits. Selenite crystals result from the reaction of sulphuric acid, produced during the decomposition of iron pyrites, with calcite in clays or mudstones. Gypsum may also be formed when sulphur-rich volcanic waters react with limestones.

In Britain important gypsum deposits are worked by opencast methods from Triassic rocks in the Midlands, and gypsum plus anhydrite are worked by deep mining from seams in the core of the Wealden Dome near Mountfield in Sussex. In the USA there are significant gypsum deposits in many States and in rocks of Silurian, Carboniferous, Permian, Triassic and Jurassic ages.

The uses of gypsum are many. It is principally employed in the manufacture of boarding and as a retardant in CEMENT. It is also used as a MINERAL FILLER in the manufacture of paint, rubber, paper, cotton, plaster of Paris, food, soil conditioners and fertilizers. Additionally it is employed for dusting colliery passages, and in the smelting of nickel. When dissolved in water during brewing it improves the final product, a process known as Burtonization. Blackboard chalk is made from gypsum, not as the name suggests from the type of limestone called chalk. Gypsum production in 1973, in tonnes, was: USA, 12 428 000; Canda, 7 544 000; France, 6 159 000.

H

Hafnium (Hf). A metal produced in small quantities as a by-product during the preparation, from the mineral zircon, of ZIRCONIUM metal for nuclear reactors. Hafnium is used as control rods in reactors. It forms no independent minerals. World production in 1968 was only 55 tonnes.

Halite. ⟡SALT.

Halloysite. ⟡CLAY MINERALS.

Hardness (1). A property of water that can be demonstrated most commonly by the amount of soap required to produce suds. Hardness might be called the soap-wasting property of water. Calcium and magnesium are the main causes: the hardness value of water is derived from the sum of the calcium and magnesium ions expressed as an equivalent amount of calcium carbonate ($CaCO_3$).

The total hardness of water may be subdivided into carbonate (temporary) and non-carbonate (permanent) hardness. Carbonate hardness includes that portion of calcium and magnesium that would combine with the bicarbonate and small amount of carbonate present. This hardness can virtually be removed by boiling which precipitates calcium and magnesium carbonates. Non-carbonate (permanent) hardness is the difference between total and carbonate hardness. It is caused by the amount of calcium and magnesium present that would normally combine with sulphate, nitrate and chloride ions. This type of hardness cannot be removed by boiling.

Water with a hardness value of less than 50 ppm is considered soft and hardness values between 50 and 150 ppm are not objectionable. At levels of 200 to 300 ppm hardness is very noticeable, and domestic supplies of this value are usually softened. In recent years suggestions have been made that there may be a correlation between cardiac problems and soft water areas.

Hardness (2). An important physical property in a mineral, useful in identification. It is determined by reference to Mohs' scale of mineral hardnesses: 1 talc, 2 gypsum, 3 calcite, 4 fluorspar, 5 apatite, 6 feldspar, 7 quartz, 8 topaz, 9 corundum, 10 diamond.

Heavy-spar. ⟡BARIUM MINERALS.

Hematite. Iron oxide (Fe_2O_3). It may occur as earthy masses or in a variety of crystalline forms. Brownish-red, steel-grey or black in colour, it may be distinguished most readily by its indian-red streak. Hematite is the commonest and most important iron ore; it contains 70 per cent iron. (⟡BANDED IRON FORMATIONS.)

Hemimorphite. ⟡ZINC.

Hinge zones. ⟡FOLD.

Hornblende. ⟡AMPHIBOLE.

Hornfels. Hard, brittle, fine-grained rock produced by thermal METAMORPHISM of a clayey rock or a LIMESTONE. Hornfels are found in the vicinity of large igneous intrusions where they may be useful sources of ROAD METAL and other heavy ballast. The quarries at Meldon near Dartmoor in Devon are big producers of hornfels track AGGREGATE for British Rail.

Hydraulic conductivity (permeability). The ease with which water moves through an aquifer, i.e. the rate of flow of ground water through unit cross-sectional area of an aquifer under unit hydraulic gradient at a fixed temperature. It can be measured in feet per second, centimetres per second or inches per day, but commonly in the analysis of ground water flow hydraulic conductivity is measured in gallons per day per square foot or metres per day. Hydraulic conductivity is noted as k and possesses a wide range in values from clays ($k = 10^{-5}$ to 10^{-7} metres per day) to gravels ($k = 1$ to 10^3 metres per day or more).

Hydraulic conductivity is sometimes referred to as coefficient of permeability, transmission constant or, simply, permeability, but because of the analogy to electrical and thermal conductivity it is best referred to as hydraulic conductivity.

Hydraulic conductivity varies according to both the characteristics of the porous medium (aquifer) and the fluid properties. In this context it is to be distinguished from intrinsic permeability which defines the pervious properties of the porous medium to the exclusion of the fluid properties.

Hydraulic continuity. When two permeable bodies adjoin and ground water can freely move from one into the other the two bodies are said to be in hydraulic continuity. For example gravels, immediately adjacent to a river channel, may lie upon a bedrock LIMESTONE AQUIFER. It is probable that the limestone and gravel would possess a mutual WATER TABLE and thus would be in hydraulic continuity.

Hydraulic fracturing. Used since the 1940s in the oil industry for increasing the yield from RESERVOIR ROCKS. To a lesser extent it is also used to increase ground water yields. If water at an abnormally high pressure is pumped underground via a borehole it will open cracks, such as JOINTS, provided that the water pressure exceeds that in the rocks which is tending to close the cracks. Thus the effective permeability of the rocks is increased, and the flow of desired fluids is facilitated.

In order to jack open the cracks, sand is pumped underground with the water which acts as a carrier for the sand. When the sand enters the opened cracks they are unable to close after the water pressure has been allowed to drop. Sand suitable for this purpose is called sandfrac sand; it is generally composed of rounded quartz grains.

Hydraulic gradient. Hydraulic head, relating to water in an aquifer, is the energy per unit weight of water. Hydraulic gradient is implicit in DARCY'S LAW where the difference in hydraulic head divided by the distance along the flow path in fluid flow is the hydraulic gradient. Thus ground water moves through an aquifer from a high inflow head to a lower outflow head, that is, in the direction of the hydraulic gradient.

A contoured water table map of an UNCONFINED AQUIFER provides a two-dimensional picture of variations in the hydraulic gradient. Where the contours are close together the hydraulic gradient is steep, which reflects low permeability of the aquifer.

Hydrocarbons. Compounds of hydrogen and carbon, some with minor or trace quantities of oxygen, nitrogen, sulphur and other elements. There are many groups of different hydrocarbon compounds but the three commonest in the FOSSIL FUELS are the paraffins, napthenes and benzene-type compounds. Some are solids, others liquids or gases.

Hydrogen ion concentration. ◊ pH.

Hydrogeology. The study of subsurface water with particular emphasis placed on its chemistry, direction of movement and relationship to the geological environment. Geohydrology differs from hydrogeology in that its emphasis is placed on the physics of flow through porous media rather than on the overall geological context of subsurface water. (◊INTERSTICES; PORES; POROSITY.)

Hydrographs. Graphical representations of either surface stream discharges or water level fluctuations in wells. The former type of hydrograph portrays the characteristics of the flow of a stream. It thus illustrates the fluctuations in stream flow in chronological order. Discharge is plotted vertically and time horizontally on the hydrograph (Figure 31).

Hydrographs of water levels in wells can be computed from measurements of water depth below ground surface. They are among the more important sources of hydrogeological information. Long-term records may allow estimates of the ultimate yield of the aquifer and the rate of its replenishment. Short-term well hydrographs, in addition to providing valuable information on the aquifer itself, have also been used to illustrate earthquake phenomena, testing of nuclear devices and sudden changes in barometric conditions.

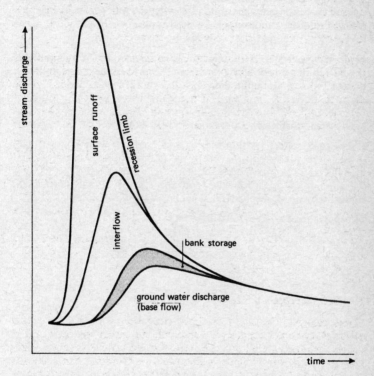

Figure 31. Hydrograph illustrating the various components of stream discharge.

Hydrological cycle. The earth's water cycle: the continuous circulation of moisture and water on our planet (Figure 3). It includes PRECIPITATION, i.e. all forms of water arriving on the earth's surface and derived from atmospheric vapour. The principal forms of precipitation are rain, mist, hail, snow and sleet, and following on from precipitation three major subsequent phases are distinguishable: EVAPORATION, surface RUNOFF and INFILTRATION. The hydrological cycle is therefore a continuous phenomenon but artificial conditions can modify the cycle, e.g. reduction of evaporation from RESERVOIRS by treating the water with cetyl alcohol or inducing precipitation through cloud-seeding using dry ice. (⬦ ARTIFICIAL RECHARGE.)

Hydrology. The study of atmospheric, surface and subsurface waters and their connection with the HYDROLOGICAL CYCLE. Thus, it is the science of water from the time that it is precipitated until it is discharged into the sea or returned to the atmosphere.

Hydrosphere. The body of (liquid) water on or near the surface of the earth. The oceans are the largest single component in the hydrosphere, cover over 70 per cent of the earth's surface and have an average depth of nearly two and a half miles.

Hydrothermal stage. ⬦ MAGMA.

Hygroscopic coefficient. ⬦ HYGROSCOPIC WATER.

Hygroscopic water. A thin film of water adsorbed by soil particles in the SOIL WATER ZONE where the adhesive forces are large and thus the water is unavailable to plants. The hygroscopic coefficient is the amount of adsorbed water on the surface of soil particles in contact with an atmosphere of 50 per cent relative humidity at 25°C.

Hypabyssal rocks. ⬦ IGNEOUS ROCKS.

I

Iceland Spar. ▷CALCITE.

Igneous rocks. Rocks formed by the crystallization or vitrification of molten rock material called MAGMA. Magma chambers occur, or occurred, at depth in parts of the earth's crust and upper mantle. Magma which is emplaced in other rocks beneath the earth's surface cools and eventually solidifies to give intrusive igneous rocks, while that erupted as lava on the surface forms volcanoes. Many intrusions are exposed at the surface today as a result of denudation which has removed the former cover of surrounding, or country, rocks. The intrusion and extrusion of magma are commonly related to relative motions between lithospheric plates (▷THE EARTH).

Magma which cools slowly at depth in major intrusions, called BATHOLITHS, gives rise to coarse-grained, plutonic igneous rocks. Magma intruded in DYKES or SILLS generally cools more rapidly, hence the size of the crystals in the resulting rock is generally medium, and these are the hypabyssal rocks. Fine-grained igneous rocks are those which have cooled rapidly generally as a result of having been erupted from volcanoes; they are the volcanic, or extrusive igneous rocks. Some volcanic rocks are glassy as a result of very rapid cooling.

Igneous rocks are classified according to the average size of the grains which they contain, and according to their chemistry, as reflected in their mineralogy. The chemical divisions are based on the percentage of silica (SiO_2) in the whole rock. The traditional groups are acid, where SiO_2 exceeds about 66 per cent; intermediate where SiO_2 is between 52 and 66 per cent; basic where SiO_2 is between about 45 and

A classification of igneous rocks

Grain size	Acid	Intermediate	Basic	Ultrabasic
Coarse >2 mm	granite	syenite/diorite	gabbro	peridotite/dunite
Medium 0·02–2·00 mm	felsite/micro-granite	porphyrite	dolerite	
Fine <0·02 mm	rhyolite	trachyte/andesite	basalt	

52 per cent; and ultrabasic, or ultramafic, where SiO_2 is less than 45 per cent. The acid rocks generally contain quartz, alkali feldspar and mica, while the basic rocks contain calcic feldspars plus augite (or other PYROXENES) and olivine. Ultrabasic rocks generally contain olivine or augite as dominant minerals. The table sets out an elementary classification of some of the igneous rocks.

The pale-coloured minerals, such as quartz and feldspar, are often referred to collectively as felsic minerals, while dark-coloured minerals, such as augite and olivine, are regarded as mafic minerals.

Ignimbrite. ▷PYROCLASTIC ROCKS.

Illite. ▷CLAY MINERALS.

Ilmenite. ▷TITANIUM.

Indium (In). A rare metallic element (crustal abundance 0·1 ppm) which is used mainly in the electronics industry and in the production of solders and alloys. It is recovered commercially solely as a by-product during the processing of zinc ores. World production in 1974 was estimated at 1·8 million troy ounces.

Induced recharge. A method of withdrawing ground water at strategic points to induce natural recharge. If ground water is pumped from an aquifer which is close to, and in HYDRAULIC CONTINUITY with, a river or lake, then the lowering of the ground water level induces additional flow of water to enter the aquifer from the surface source. COLLECTOR WELLS may be regarded as items of induced recharge (Figure 17).

Infiltration. The vertical flow of water through the soil surface into subsurface materials. For water to infiltrate the soil surface must be suitable. Once infiltration takes place after soil moisture deficits have been satisfied, movement of the water is essentially downwards through the AERATION ZONE towards the WATER TABLE. However, there may be some lateral movement of water in this zone such as INTERFLOW. Infiltration should be distinguished from PERCOLATION, the former pertaining to vertical downward movement of water in the zone of aeration and the latter to flow of ground water in mainly lateral directions within the ZONE OF SATURATION.

Infiltration amounts may be measured by infiltrometers comprising apparatus in which the rate of intake is determined directly as the rate at which water must be added to maintain a constant depth within the infiltrometer.

Infiltration galleries. Horizontal, permeable or hollow conduits for intercepting and collecting ground water by gravity flow. These galleries, many of which date from ancient times, are normally constructed at no great depth below ground surface and to work efficently they should be in aquifer with a high WATER TABLE fed by an adequate nearby water source of suitable quality. Some galleries are deeper and are driven as headings from wells many feet below the surface to intercept the maximum flow path of ground water in the ZONE OF SATURATION. (◊KANATS.)

Influent stream. A stream which loses water to the ground water body and thus contributes to the ground water reserves (Figure 23). The local build-up of head on the WATER TABLE underlying an influent stream is termed a ground water mound. (◊EFFLUENT STREAM; INTERMITTENT STREAM.)

Interflow. Interflow is often regarded as subsurface RUNOFF. During INFILTRATION some water may move laterally at the soil–bedrock interface and contribute to surface runoff in permeable catchments. The water that moves through the ground as interflow will appear in the stream prior to the main BASE FLOW or ground water component from the aquifer (Figure 7). In this context the arrival and amount of interflow can be computed from an analysis of the river HYDROGRAPH from a permeable catchment.

Intermediate rocks. ◊IGNEOUS ROCKS.

Intermediate zone. A zone or subzone which occurs within the AERATION ZONE (Figure 7) and extends between the SOIL WATER ZONE and the CAPILLARY FRINGE. It is variable in thickness from near zero to several hundred feet in arid areas. Liquid and gaseous phases are present within this zone. Non-moving or PELLICULAR WATER in the intermediate zone is held there by hygroscopic and capillary forces. Excess water is GRAVITATIONAL WATER which moves downward, on a near-vertical path, under the influence of gravity.

Intermittent stream. A stream which, in general, flows during wet seasons but not during dry seasons. The WATER TABLE lies above the stream bed during the wet season but falls below it during the dry season. Thus the flow is derived principally from surface RUNOFF but during wet seasons also receives some ground water body. Intermittent streams are, therefore, both EFFLUENT and INFLUENT in nature depending upon the time of the year.

Interstices. The portions of rocks or soils which are not occupied by solid matter; also, VOIDS or PORES. Interstices may comprise small spaces between rock matrix particles but are of fundamental importance in HYDROGEOLOGY as they can act as ground water conduits. They are characterized by their size, shape, irregularity and distribution.

Original interstitial space is produced by the processes leading to the formation of particular rock types. Secondary interstices, developed after the rock was formed, are due mainly to stresses imparted to that rock. Examples include JOINTS, FAULTS and solution openings.

From the point of view of size, interstices may be classed as capillary, supercapillary and subcapillary. Capillary interstices are sufficiently small so that surface tension forces will hold water within them. Supercapillary interstices are larger while subcapillary ones are so small that water is retained within them by strong, adhesive forces. If there is good connection between the interstices they are said to be of communicating type whereas bad connection will result in isolated interstices.

All subsurface water contained within interstices, both in the AERATION ZONE and the ZONE OF SATURATION, is called interstitial water.

Intrinsic permeability. Intrinsic permeability defines the pervious properties of a porous medium (aquifer) to the exclusion of the fluid properties. Thus intrinsic permeability comprises a part of HYDRAULIC CONDUCTIVITY.

Intrusive rocks. ◊IGNEOUS ROCKS.

Iodine sources. A major proportion of the world's iodine (I) is obtained as a by-product of the working of the Chilean NITRATE DEPOSITS. Iodine is a non-metallic element belonging to the halide group. The caliche, or host rock, containing the nitrates possesses an average iodine content of 0·04 per cent in the form of the iodate minerals lautarite ($Ca(IO_3)_2$) and detzeite (calcium iodate plus calcium chromate). Although seawater contains only 0·05 ppm of iodine, some seaweeds can concentrate it in their cells up to 0·45 per cent dry weight. After burning, the kelp, or ash, contains 1·4–1·8 per cent iodine. Iodine has been extracted from seaweed in Eire, France, Japan, Norway, Russia, the UK and the USA. Some iodine is also obtained from NATURAL GAS and oil-well brines.

Iodine is used in several chemical industries, in the processing of animal feeds, in medicine and in photography.

Ionic concentration. In a chemical analysis of ground water, concentrations of different ions are expressed by weight or by chemical equivalence. A complete chemical analysis of a ground water sample includes determinations of the concentrations of all the inorganic constituents present. Dissolved salts in ground water occur as dissociated ions: the common cations and anions present include calcium, magnesium, sodium, potassium, carbonate, bicarbonate, chloride, sulphate and nitrate.

The dissolved salts in ground water affect its usefulness for various purposes. If one or more of the ions is in excess of the amount that can be tolerated for a given use, some form of treatment may be applied to change or remove the undesirable salt so that the water will serve for the intended purpose.

Iridium. ⟡PLATINUM-GROUP METALS.

Iron (Fe). The twentieth century may be designated the 'Space Age' or the 'Nuclear Age' but in terms of our dependence on a single metal it is still the 'Iron Age'. Iron and its various alloys (steel) are indispensable to a modern industrial civilization. It is the second most abundant metal (5·8 per cent) in the earth's crust, and it accounts for more than 95 per cent of all metals consumed. Further, large tonnages of other rocks (coal and limestone) and minerals (⟡FERROUS MINERALS) are mined principally to facilitate the steel-making process or specifically to modify the end product.

Because of its crustal abundance, there are many minerals which contain iron, but the most important as ores are the oxides (hematite, magnetite and goethite) and the carbonate (siderite). These minerals occur in a wide variety of geological environments and at many geographic localities. Deposits of chemical-sedimentary origin (i.e. deposited in lakes or seas as chemical precipitates) are the commonest type of ore. IRONSTONES were important in the past but, increasingly, the BANDED IRON FORMATIONS form the bulk of present-day production.

Total world production of iron ore (1974) was 810 million tonnes. Over one-quarter of this came from the USSR. Other countries producing more than 50 million tonnes were the USA, Australia, France, Brazil and Canada. Over forty countries produced significant (more than 0·5 million tonnes) quantities of iron ore. The United Kingdom production, mainly from low-grade Jurassic ironstones, totalled 3·6 million tonnes.

Ironstones. Bedded sedimentary rocks containing notable amounts (typically 20–40 per cent) of IRON. A wide variety of iron minerals may be present but one or more of geothite (oxide), chamosite (silicate) and siderite (carbonate) are common. Oolitic, pelletal and sometimes pisolitic structures are typical.

Ironstones are predominantly of post-Precambrian age; in Britain (Frodingham, Northampton and Cleveland) they are concentrated in the Jurassic system. Synonyms include iron ores of the 'minette' or 'Lorraine' type. Although economically important in the past and still worked today, they are now of less importance than the BANDED IRON FORMATIONS. (⟡DIMENSION STONE.)

Irrigation. The artificial watering of fields for crop production. The suitability of a water source for irrigation depends upon the effects of the mineral content of the water on both the crop and soil. Specific limits of permissible salt concentrations for irrigation water are difficult to state because of the wide variations in salinity tolerance among different plants. (◊ARTIFICIAL RECHARGE; WATER SPREADING.)

Itabarite.   ◊BANDED IRON FORMATIONS.

J

Joint. A joint in a rock mass is a barren, closed FRACTURE plane along which there is evidence of there having been slip. Most consolidated rocks are cut by many joints. Parallel joints form a joint set; usually several sets cut a rock mass. Many joints in sedimentary rocks are thought to be formed by the release of residual strain energy when the rocks are brittle. Some joints in igneous rocks are related to shrinking during cooling. In some basalt lava flows, DYKES and SILLS, joints of this type bound long columns.

Joints divide many rock masses into rectangular blocks. Where joints are closely spaced they render the rock unsuitable for use as DIMENSION STONE.

The movement of ground water in crystalline or well-cemented rocks is largely by flow along fissure planes. As it is commonly understood, a fissure is an open or dilated joint or other fracture plane. In engineering geology the name fissure is given to small, commonly curving, joints.

Juvenile water. Primary or new water which has not been part of the HYDROLOGICAL CYCLE before and has been derived mainly from the interior of the earth. Juvenile water may be of magmatic, volcanic or cosmic origin, the latter being distinguishable in that it contains the gas argon. (⟡MAGMATIC WATER.)

K

Kainite. ⬦POTASH DEPOSITS.

Kanats. Long INFILTRATION GALLERIES which collect water from alluvial deposits or other sedimentary rock aquifers. Kanats, which collect water for both domestic and agricultural purposes, were probably first used more than 2500 years ago in Iran. One extensive kanat system built about 500 BC in Egypt is said to have irrigated 4700 square kilometres west of the Nile. Many kanats are still used as water sources today in Iran and Afghanistan.

Kaolin. ⬦CLAY MINERALS.

Kaolinite. ⬦CLAY MINERALS.

Karst. An area underlain by limestone strata marked by very large solution openings. These permeable regions often have characteristic undulating surfaces with conical knolls and circular sinkholes or swallets. The term karst topography is applied to such areas.

Thus a karst area comprises an UNCONFINED AQUIFER where RUNOFF usually enters the ground through sinkholes and fissures and pursues a course to an outlet (spring or rising) through a system of underground passages or solution channels. Drainage in such an area may be entirely underground and, although complex, somewhat similar to a surface drainage net in its characteristics. Extensive karst regions occur in France, Yugoslavia and the USA.

Kernite. ⬦BORATE DEPOSITS; SODIUM CARBONATE AND SULPHATE MINERALS.

Kieselgur. ⬦DIATOMITE.

Kyanite. ⬦SILLIMANITE GROUP OF MINERALS.

L

Laminar flow. A fluid flow regime where the fluid moves with a relatively slow motion in which each 'thread' of liquid retains its identity and flows smoothly alongside its neighbours. It is a type of streamline motion which curves smoothly around irregularities in its path rather than setting up whirls and eddies as it moves past.

Laminar flow conditions in fluid flow through aquifers is implicit in DARCY'S LAW. High flow velocities, particularly encountered in the immediate vicinity of pumping wells discharging at high abstraction rates, will produce turbulent conditions, a departure from linearity, and Darcy's Law breaks down.

LANDSAT. Orbiting satellites sent up by the American NASA to collect imagery from the earth for purposes of surveying its geological, hydrological and biological surface features. Two satellites have been put into orbit. LANDSAT 1 has been in orbit nearly three and a half years (November 1975) and is showing signs of wear. In three years of life the satellite circled the earth some 15270 times and sent many thousands of messages and images to receiving stations in Europe and the Americas. LANDSAT 2 will take over the functions of its predecessor as the original equipment begins to fail. (◊ERTS; REMOTE SENSING.)

Langbeinite. ◊POTASH DEPOSITS.

Larvikite. ◊SYENITE.

Laterite. A residual deposit, formed by the chemical weathering of rocks, which is extensively developed in humid, tropical and subtropical regions. It is leached of silica and elements such as sodium, potassium, calcium and magnesium; iron and aluminium hydroxides commonly remain. If iron predominates, the deposit is termed a laterite; if (more rarely) aluminium predominates the deposit is termed a BAUXITE.

Other metals, notably nickel, cobalt and chromium, may be concentrated in sufficient quantities in laterites so as to form low-grade – but very extensive – ore deposits. The important nickeliferous laterites of New Caledonia are of this type.

Lautarite. ◊IODINE SOURCES.

Lava. ◊MAGMA.

Lead (Pb). One of the earliest metals used by man; the hanging gardens of Babylon had lead pans to hold plants and the Romans used it extensively as water piping. The major uses today are in storage batteries (where much lead is recycled) and as lead tetraethyl for adding to petrol, to improve engine performance in motor cars. Engine performance is certainly improved but it is arguable whether the resulting air pollution (and permanent loss of lead) justifies the improvement. Other diverse uses include the construction industry, in pigments, as a component of solders for tin cans, in the manufacture of small-arms ammunition and as printers' type metal.

The principal ore of lead (crustal abundance 13 ppm) is GALENA which is commonly associated with ZINC (sphalerite) and other BASE METAL sulphides. Lead and zinc ores occur as veins associated with igneous intrusions, as STRATIFORM ORE DEPOSITS and as replacement deposits in limestones. The limestone–lead–zinc association of the latter represents one of the world's major sources of lead and zinc; it is variously known as 'Alpine', 'Mississippi Valley' or 'Tri-state', and 'Silesian' type of mineralization.

World mine production in 1974 was an estimated 2·7 million tonnes (metal content of lead) and came from mines in over forty countries. The USSR, USA, Canada and Australia produced half the world's total.

Leucite. The FELDSPATHOID MINERAL ($KAl(SiO_3)_2$) which characteristically occurs in some silica-deficient volcanic igneous rocks, and is a potential source of POTASH. Unfortunately it is rarely concentrated in sufficient quantities.

Lignite. ◊COAL.

Lime. When LIMESTONE is calcined ('burnt') in a lime kiln, carbon dioxide is driven off from the mineral CALCITE ($CaCO_3$) to yield quicklime (CaO). When water is added to quicklime a strongly exothermic reaction gives slaked or hydrated lime ($Ca(OH_2)$), a material which sets on exposure to the atmosphere. It is used for making mortar. The presence of clay in a limestone has the effect of slowing the slaking process and may allow the material to become hydraulic, that is, suitable for use as CEMENT. Calcareous fossil shells in a limestone tend to make the rock calcine unevenly; consequently slaking does not occur properly.

The term lime is also applied to finely powdered limestone which may be spread on the land in order to neutralize acid soils and to supply plant nutrients. Quicklime and slaked lime are less commonly used for the same purpose.

The principal use of lime is in the chemical industry as a low-cost alkali. Lime obtained from DOLOMITE is used as a REFRACTORY material. A most important use of lime is in the fluxing of steel and other metals, in the making of paper and ceramics, and in water and sewage treatment.

Limestone. A bedded sedimentary rock composed mainly of the mineral CALCITE ($CaCO_3$). Rocks containing calcite and DOLOMITE (($CaMg)CO_3$) are known as dolomitic limestones, and those composed primarily of the mineral dolomite are known as dolomites. The name magnesian limestone is used by some geologists for a limestone which contains calcite with some magnesium in its lattice. The limestones are among the most abundant sedimentary rocks in many countries of the world. Limestones and dolomites are commonly thought of as constituting a broad group of sedimentary rocks known collectively as the carbonates.

Most limestones were deposited beneath the sea as a result of the accumulation of fossil shell debris, or as a consequence of the precipitation of calcite. A few limestones originated on the floors of lakes, or in river systems.

CHALK is the name given to a soft, pure, white, fine-grained type of limestone. Coquina is a soft, porous limestone composed of shell and coral debris. Along the south-eastern coast of the USA it is used for CEMENT making and secondary road building (⬦ROAD METAL). Oolitic limestones or oolites are composed of spherical grains of calcite up to about 3 millimetres in diameter. In the Cotswold Hills of England oolites of Jurassic age are much valued for DIMENSION STONE, or for making ARTIFICIAL STONE; locally they are known as Freestones. Travertine and tufa are superficial encrustations of limestone formed as a result of the precipitation of calcite from surface or ground waters. Travertine is compact, while tufa is soft and porous. Both have been used for ornamental purposes. The term MARBLE is used in the stone trade to describe any calcite-rich rock which is attractive when polished; to a geologist a marble is a metamorphosed limestone.

Most limestones are deposits of low MONETARY OR UNIT VALUE but high PLACE VALUE. The uses of limestones are many and varied as the following abbreviated list indicates: ABRASIVES; AGGREGATE; ARTIFICIAL STONE; CEMENT; DIMENSION STONE; CRUSHED STONE; DOLOMITE; GLASS; LIME; ROAD METAL. Limestone is also used as a flux in iron and steel making, in the manufacture of soda ash and calcium carbide, and in the processing of sugar and paper. Many important oil reservoirs and aquifers are limestones.

Limestone aquifers. LIMESTONE is an important type of sedimentary rock composed mainly of calcium carbonate (CALCITE). The related sedimentary carbonate rock DOLOMITE contains large amounts of

calcium–magnesium carbonate. Most limestones or dolomites originally have relatively few connected INTERSTICES. Subsequent earth
movements produce FRACTURES and JOINTS in the material. Water
enters these openings and, under certain conditions, slowly dissolves
some of the rock. This action enlarges some of the openings to form
solution channels and where a network of channels are present, such
rocks will yield copious supplies of water to individual wells.

Limestone aquifers contain important water resources in many
countries, e.g. Britain and the USA where several million gallons per
day may be pumped from wells sited in an ideal geological setting.
Some of these large yields have been produced by ACIDIZATION, to
increase the size of the solution openings, after well drilling has taken
place.

Limonite. A mixture of earthy hydrous iron oxides. Usually yellow or
yellowish-brown in colour and results from the alteration of other iron
minerals.

Lithium (Li). The lightest metal. Its compounds are used principally
in multi-purpose lubricating greases, in the GLASS and ceramic industries, as dust absorbers in air conditioners, and during the production
of ALUMINIUM.

Spodumene, petalite, amblygonite and the lithium mica lepidolite
are the source minerals; they occur in granite PEGMATITES. Lithium
compounds are also recovered by solar evaporation of natural brines.

Accurate statistics are unobtainable, but the USA dominates world
production.

Lithophone. ◊MINERAL PIGMENTS.

Lithospheric plates. ◊THE EARTH.

Lode. A miner's term for a fissure filled with a concentration of one
or more (usually metalliferous) minerals. Particularly used to describe
quartz and gold VEINS.

Log. The record of events or the type of rock penetrated in drilling
a borehole as indicated by samples or by information obtained from
electronic devices.

C. A. Moore, *Handbook of Subsurface Geology*, Harper & Row, 1963.

Lysimeter. An experimental installation to evaluate INFILTRATION
and evapotranspiration (◊TRANSPIRATION) under field or natural
conditions. The lysimeter is normally a tank, up to 2 metres square and
2 metres deep, filled with earth and embedded with the surface almost
flush to the ground. The bottom is funnel-shaped and drains into a
closed receptacle located in an underground chamber. The soil evaporation is the difference between the rainfall and the drainage.

M

Mafic minerals. ⟡IGNEOUS ROCKS.

Magma. Molten rock material within the earth's crust; cooling and solidification will give an IGNEOUS ROCK. Magma may be entirely fluid, or it may contain mineral crystals within the liquid. On eruption at the surface of the earth, and with the escape of its contained gases, magma becomes lava. If magma is intruded without reaching the surface, it forms BATHOLITHS, DYKES or SILLS. Magmas have been the sources of many extensive bodies of minerals conveyed into the surrounding rocks. They are composed of non-volatile and volatile constituents. On cooling, when the magma begins to crystallize, the anhydrous minerals are precipitated first; this forms the orthomagmatic stage. As a result of the early separation of the anhydrous minerals, there is a concentration of gases and liquids in the residual magma. The liquid part crystallizes as the pegmatitic stage and the gaseous part as the pneumatolytic stage. The final hot-water-rich solution forms the hydrothermal stage. Ore deposits are believed to be associated with each stage.

Magmatic water. Water in or derived from MAGMA. It is, therefore, a form of new or JUVENILE WATER and has never been part of the HYDROSPHERE.

Magnesia. ⟡MAGNESIUM.

Magnesian limestone. ⟡LIMESTONE.

Magnesite. The magnesium carbonate mineral magnesite ($MgCO_3$) is found in three associations. As a result of the activity of hydrothermal solutions it may replace dolomite or limestone. Veins of cryptocrystalline magnesite plus silica cut some serpentines or altered ultrabasic igneous rocks. Magnesite also occurs in association with some EVAPORITE DEPOSITS. It is thought that it was initially precipitated as hydromagnesite and converted to magnesite by dehydration.

Important deposits of magnesite are present in Manchuria, Austria, California and Nevada. Magnesite is a source of MAGNESIUM and magnesium compounds; when 'burned' it yields a REFRACTORY material. It is also used in GLASS and paper making, as an ABRASIVE, in ceramics, as a FERTILIZER, and in the making of Epsom salts. Production of crude magnesite in 1972, in tonnes, was: North Korea, 1700000; USSR, 1485000; Australia, 1429000.

Magnesium (Mg). The eighth most abundant element (2·8 per cent) in the earth's crust and the third most abundant element dissolved in seawater. It is extracted commercially from seawater and other natural brines, and from the minerals dolomite ($CaMgCO_3$), magnesite ($MgCO_3$), olivine ($(MgFe)_2SiO_4$) and brucite ($Mg(OH)_2$). The supply of magnesium is, for all practical purposes, inexhaustible.

Most magnesium is used as the oxide (magnesia) in the manufacture of metallurgical REFRACTORY bricks. Other diverse uses include the manufacture of CEMENTS used in flooring, making paper pulp, rayon flares and FERTILIZERS. In the metallic form it is used, particularly in ALUMINIUM alloys, wherever lightness is desired; the aircraft and automobile industries are major consumers. (◊LIMESTONE.)

Statistics for the production of magnesium minerals are difficult to obtain, but the UK, India, Belgium and Japan are major producers of dolomite, while the USSR, Austria, Czechoslovakia and North Korea are the major sources of magnesite. Production of magnesium metal (1973) was an estimated 265000 short tons; the USA, USSR and Norway were the major producers.

Magnesium salts. Most of the valuable magnesium salts are sulphates. The evaporite mineral (◊EVAPORITE DEPOSITS) epsomite or Epsom salt ($MgSO_4.7H_2O$) is present dissolved in seawater and in some brines. It is generally extracted by crystallization from such waters, as, for example, at Epsom in Surrey, England, where it is obtained from mineral waters. In the USA several lake and well brines in the Mid-west and West yield epsomite on evaporation.

Solid epsomite occurs in association with some salt beds, for example, in the German POTASH DEPOSITS near Stassfurt. Much Epsom salt is produced by reacting sulphuric acid with magnesite or dolomite.

Magnesium sulphate is used in the manufacture of artificial fibres and in the leather, paint, soap, paper, FERTILIZER and textile industries.

Magnetite. Iron oxide (Fe_3O_4) containing 72 per cent iron. It is a common and important ore of iron, and is characterized by its black colour, its hardness (6), and its strong magnetism.

Malachite. A bright green copper carbonate, $Cu_2CO_3(OH)_2$, found in the zone of oxidized enrichment (◊SECONDARY ENRICHMENT), associated with other copper minerals. An ore of copper; the mineral contains 57·4 per cent copper.

Manganese (Mn). The twelfth commonest element, widely distributed and occurring in a large number of minerals and a wide range of geological environments. Commercial deposits, however, are unevenly distributed and are predominantly of chemical-sedimentary or of residual origin. The commonest minerals are oxides (PYROLUSITE) and often earthy, complex hydrated oxides (e.g. PSILOMELANE).

Manganese is essential for the production of cast iron and steel where it is used to counteract the harmful effects of sulphur and oxygen, and as an additive in certain alloys. With present technology there is no substitute. Manganese dioxide is used in dry-cell batteries.

The metal is of considerable strategic importance, since few of the leading Western industrial nations have sufficient supplies. About half the world's production is in the Communist bloc, particularly from the Nikopol basin in the Ukraine and the Chiatura basin in Georgia in the USSR. Other important producers are South Africa, Brazil, Gabon and India. Total world production in 1974 was an estimated 25 million tonnes.

Manganese nodules. Brownish-black nodules (1–30 centimetres in diameter), composed of earthy MANGANESE and IRON oxide–hydroxide compounds, found on the sea bed in all the oceans. In the Pacific alone, it is estimated that tens of millions of square miles are covered with these nodules. Although variable in composition, some contain up to 41 per cent manganese and important amounts of COPPER, COBALT and NICKEL; they may represent one of the earth's great untapped sources of these metals. Several countries are now actively investigating raising them from the sea bed. Some writers suggest they might be accumulating certain metals faster than man consumes them – offering the intriguing possibility of a mineral deposit growing faster than it can be mined.

Technological considerations aside, there are international legal problems of ownership and mining rights of minerals on the deep sea ocean floor.

Manganite. Hydrated manganese oxide ($MnO(OH)$), characterized by its black colour, prismatic crystals, hardness (4) and brown streak. A minor ore of manganese.

Mantle. ◊ THE EARTH.

Marble. Recrystallized LIMESTONE, thus a metamorphosed rock of which CALCITE forms the greater part and other minerals in minor quantity give the colour and other decorative features. Pure marble is white. Most marbles worked for sculpture or architectural purposes are massive and should show little or no preferred direction of parting or fracture. The stone has been known and worked since the earliest days of architecture in the Mediterranean, Near East and Far-Eastern areas. Famous quarries yielding vast quantities of stone in pre-Christian times were excavated in Greece and Italy, and several are still in production. Other famous quarries occur in Vermont and New York States. Little true marble has ever been produced in Britain, but metamorphosed limestone from South Devon has from time to time been in demand for ornamental purposes. ($\Diamond$ DIMENSION STONE.)

Marl and marlstone. Marl is a soft, relatively unconsolidated rock comprising a mixture of CLAY or silt and fine-grained aragonite or CALCITE mud. The clay or silt fraction of the rock should not exceed about 70 per cent, or be less than about 30 per cent. When wet, marls are plastic, but when dry they are friable. Marlstone is the name given to a rock of marl composition when it is consolidated. In Britain many rock units named as marl, e.g. the Keuper marl of Triassic age, contain little true marl according to the definition given above. The term marl is also used elsewhere in the world for some soft, friable limestones.

Marls and marlstones are of relatively little value except in the manufacture of CEMENT.

Meerschaum. A soft mineral ($H_4Mg_2Si_3O_{10}$) which floats on water when it is dry, but is plastic when wet (also called sepiolite). Meerschaum is found as nodular masses in some altered magnesium-rich rocks such as serpentine and magnesite. The most famous deposits occur in Turkey where meerschaum has been exploited for at least 2000 years. Meerschaum has also been worked in Kenya, Tanzania and South Africa, and is known to occur elsewhere in the world.

Block meerschaum is carved into good-quality pipes for smokers. Compressed meerschaum dust and chips can also be used for somewhat poorer-quality pipes.

Mercury (Hg). A scarce element (crustal abundance 0·08 ppm) and the only one that is liquid at ordinary temperatures. This property, together with its high SPECIFIC GRAVITY and electrical conductivity, gives mercury a unique position in industry. It is used principally in the electrical industry (including the mercury battery) and in the manufacture of chlorine and caustic soda. Other uses include mechanical measuring devices, in fungicides and bactericides, agricultural chemicals and pharmaceutical preparations.

Man-made pollution (with tragic results in Japan) has emphasized the toxic qualities of mercury and its compounds. Recent work, however, suggests that volcanic activity may also be a significant contributor to pollution of the environment by mercury.

Globules of native mercury do occur in nature but cinnabar (HgS) is the only commercial ore. Mercury is bought and sold in strong, wrought-iron flasks which contain 76 lb (34·5 kilogrammes) of mercury. Estimated world production in 1974 was 262000 flasks; Spain, the USSR, Italy and Mexico were the major producers.

Metallogenic epoch. A geological period during which notable ore formation took place. Thus the Precambrian BANDED IRON FORMATIONS which are present in every continent show a remarkable cluster around an 'iron epoch' 2200 million years ago.

Metallogenic province. Defined and elaborated differently by different geologists, but essentially a geographic or structural region characterized by a marked concentration of ore formations. The ore formation may have taken place during a single METALLOGENIC EPOCH or in several distinct epochs.

Notable examples are the Zambian–Katangan copper belt (copper, lead and zinc), the Western Malaysian tin belt (tin and tungsten) and, on a global scale, the porphyry copper province of the Rocky and Andes Mountains.

Metamorphic rocks. ◊METAMORPHISM.

Metamorphism. Change in the character of a rock, commonly a chemical or mineralogical change brought about by temperature rise, rise in pressure, or both. Where change in the rock substance is caused by the removal of some chemical constituents or the introduction of new ones the term metasomatism is used. Metamorphism may give rise to minerals of commercial value such as garnet and talc.

Metasomatism. ◊METAMORPHISM.

Meteoric water. A high percentage of all surface and ground waters are meteoric in origin, i.e. derived from PRECIPITATION. Meteoric waters, therefore, are those which have been recently involved in atmospheric circulation.

Mica. Many of the minerals in the mica group can be cleaved into exceptionally thin, flexible and elastic sheets which are commonly transparent or translucent, and infusible. Other valuable properties of some micas include their low thermal conductivity and high dielectric strength. Mica is regarded as a STRATEGIC MINERAL because of its uses in the electrical industry.

Two micas of commercial significance are the white mica, muscovite $((H,K)AlSiO_4)$, and the amber mica, phlogopite $(H_2KMg_3Al(SiO_4)_3)$. The dark mica, biotite $(H_2K(Mg,Fe)_3Al(SiO_4)_3)$, is of little economic value. The LITHIUM mica, lepidolite, is one source of lithium.

Muscovite, economically the most important mica, occurs in commercial quantities in some granite pegmatites. Some of the mica crystals in these pegmatites are very large, up to a metre across, and are sometimes called 'books' of mica because the mineral is so readily split along its basal CLEAVAGE. The pegmatites are thought to have formed by crystallization from late-stage granitic fluids enriched in volatiles, and intruded into cracks in the main mass of a granite or the country rocks surrounding it. Perhaps the best-known pegmatitic source of muscovite is in the Bihar district north-west of Calcutta. About 600 small mines are worked in pegmatites cutting Archaen gneisses in the Indian Shield. Muscovite is also worked near Madras and Rajputana, again from Precambrian rocks in the Indian Shield. Mica pegmatites are also worked in southern Brazil.

Deposits of phlogopite are mainly associated with some coarse-grained, ultrabasic igneous rocks intruded into metamorphosed lime-stones or gneisses. Its sporadic occurrence makes it difficult to exploit. Some, however, is worked from ultrabasic pegmatites in Ontario and Quebec, and in Madagascar.

High-grade sheet muscovite, mainly from India, is used in the manufacture of electrical condensers, vacuum tubes, telephones, dynamos and for various insulation purposes. Phlogopite is principally employed in the making of commutators. Some of the waste from the processing of sheet mica, and mica from lower-grade deposits, is used in paints, as a filler ($\diamond$MINERAL FILLERS) and as a decorative agent for roofing and paving material. The 'glitter' on Christmas cards is commonly muscovite. Attention is being given to finding uses for much of the fine-grained flake mica recovered from the processing of the CHINA CLAY deposits associated with the granites of south-west England. In former times sheet mica was used for stove windows. Mica production in 1971, in kilogrammes, was: USA, 115296000; India, 18562000; South Africa, 7162966. ($\diamond$GRANULES.)

Microline. $\diamond$FELDSPAR.

Millstone grit. $\diamond$GRIT AND GRITSTONE.

Mineral. A naturally occurring solid, substance having a definite, homogeneous chemical composition. Nearly all minerals are crystalline. The word is also loosely used to denote any material obtained by mining. MINERAL DEPOSITS are thus any masses of minerals or any rock which may be worked for the recovery of an economic mineral or metal.

C. Rogers, *Rocks and Minerals*, Ward Lock, 1973.
F. H. Plough, *A Field Guide to Rocks and Minerals*, Constable, 1970.
J. Watson, *Rocks and Minerals*, Allen & Unwin, 1972.
W. R. Jones, *Minerals in Industry*, Penguin, 1963.

Mineral and rock wool. Fibrous artificial silicate glass is known as mineral wool in the USA and as glass silk and silicate cotton in Europe. Mineral wool manufactured from slag is often called rock wool. Natural rocks which have been used for the manufacture of mineral wool include sandy dolomitic limestones and calcareous mudstones. It may also be made from a mixture of limestone and silica derived from either sand or silica waste from ceramic industries. Potentially suitable minerals are WOLLASTONITE, tremolite and hornblende, the latter two belonging to the AMPHIBOLE group.

The principal use of mineral wool is in thermal and sound insulation where it has for some purposes replaced diatomite and asbestos.

Mineral deposit. A natural concentration of a MINERAL or minerals. Although the word mineral may have a precise scientific or legal connotation, in everyday usage it is commonly used to cover all non-living, naturally occurring materials that are useful to man.

Mineral deposits thus include not only the metalliferous ores, but also non-metalliferous materials (e.g. aggregate), fossil fuels such as petroleum and natural gas, as well as water and the gases of the atmosphere.

Many non-metalliferous 'minerals' (e.g. coal, aggregate) and some metalliferous ores (e.g. ironstones) are sedimentary rocks; their origin is bound up with observable and, for the most part, understandable processes which take place on the surface of the earth. A large number of metalliferous minerals, however, appear to be related to processes which emanate within the earth's crust and which are related to the crystallization of igneous rocks from magmas.

Current research has emphasized the close relationship between igneous (especially volcanic) activity and the genesis of many metalliferous ore deposits. There is, however, considerable evidence that, in many cases, the metals of the ore fluids are not derived from the magma itself but have been leached from the surrounding rock into which the magma has been intruded.

Mineral exploration. The miners' adage that 'gold is where you find it' may still apply for those, still considerable, unexplored areas of the earth. Each year, however, the chances of finding valuable minerals on the surface of the earth become more slender. Increasingly, our future mineral supplies depend on the discovery of concealed, subsurface ore bodies.

Modern mineral exploration may involve the systematic investigation of, at first, large areas using techniques such as REMOTE SENSING (◊ERTS), aerial photography and regional geological surveys, then more detailed prospecting of smaller areas. GEOPHYSICAL and GEOCHEMICAL surveys are of considerable importance. Areas of possible mineral concentration are evaluated by DRILLING.

In addition to overcoming the natural odds against finding an economic mineral deposit, the present-day prospector has to deal with objections from the 'conservationists', legal problems of the ownership of mineral rights, and taxation by governments. Notwithstanding the many obstacles, it is still possible for the lone prospector to locate a new ore body.

Mineral fillers. Many minerals and materials derived from them are used as fillers in a wide variety of industrial products. Suitable minerals are chemically inert with respect to the material they are mixed with. The principal products in which mineral fillers are used are paper, paint, plastic, rubber, pesticides, fertilizers, rubber and bitumen. Among the many rocks or minerals used as fillers are asbestos, barite, clay (◊CLAYS AS FILLERS), diatomite, feldspar, gypsum, limestone, mica, perlite, pumice, pyrophyllite, silica, slate, tripoli, vermiculite and wollastonite.

Mineral pigments. Some minerals and rocks are used after relatively simple processing to impart colour or opacity to materials such as paint, plaster, cement or rubber. Iron oxide minerals such as hematite and limonite can be used to produce red or brown colours. Among the earliest-used mineral pigments were various iron- or manganese-stained clays such as ochres, siennas and umbers giving yellow, orange and brown colours respectively. Barite is employed to produce lithophone for the manufacture of white paint.

Mineral spring. A natural spring discharge, the water possessing a high MINERAL content. A mineral water is one having a total dissolved solid content of more than 1000 milligrammes per cubic decimetre. Where the concentration is more than 100000 milligrammes per cubic decimetre the water is classed as a natural BRINE. A thermal water, commonly associated with mineral springs, is one at a temperature higher than the average mean annual air temperature on discharge at the surface. The majority of waters described as mineral and thermal are spring waters, or well waters originating from depths less than 30 metres, nearly all of which have at one time been used for medicinal or industrial purposes.

Mining. The extraction of minerals from the earth. A mine may be a surface (OPENCAST) or an underground excavation. A distinction, usually generally understood but difficult to define, is drawn between a quarry and a surface mine; thus a limestone quarry but an opencast coal mine. An underground mine implies men working beneath the ground, so petroleum, brine and sulphur wells are generally excluded.

Mining of unconsolidated deposits covered by water is accomplished by DREDGING. Novel methods of mining, which may become important in the future include bacterial leach mining (◊BACTERIAL EXTRACTION OF METALS).

Mirabilite. ◊SODIUM CARBONATE AND SULPHATE MINERALS.

Moisture equivalent. The amount of water which a saturated soil will retain after being centrifuged at a centrifugal force 1000 times that of gravity. It is a form of specific yield of the material. The FIELD CAPACITY for sands is higher than the moisture equivalent, but about the same for calcareous clays.

Molybdenite. ⟡MOLYBDENUM.

Molybdenum (Mb). A rare element with a crustal abundance of 5 ppm. Currently an important additive used in the production of stainless and alloy steels, particularly the high-strength and high-temperature types. In the future, it is likely to be increasingly in demand by the nuclear and space-age technologies; the command module of the Apollo spacecraft was covered with a stainless steel alloy containing 8 per cent molybdenum.

The main primary ore is molybdenite but substantial quantities of the element are recovered as a by-product during the processing of other, especially copper, ores. The principal reserves occur in the USA (primarily molybdenite), Chile (by-product of copper ores), and the USSR (by-product of copper ores). World production (metal content) in 1974 was 160·5 million lb.

Monazite. ⟡RARE EARTH ELEMENTS.

Monetary or unit value. Deposits of minerals of high monetary or unit value fetch high prices per tonne. Because most of them are rare, they are the minerals which figure prominently in world trade. Most of the metalliferous deposits fall into this category, as do many of the NON-METALLIC ROCKS AND MINERALS with the exception of those used in the construction industry, which are mainly of low unit value, but of high PLACE VALUE.

Montmorillonite. ⟡CLAY MINERALS.

Mud and mudstone. Mud is a moist or wet, loose mixture of particles of less than $\frac{1}{16}$ millimetre in diameter. If mud consists of particles which are mainly less than 1/256 millimetre in diameter it is called CLAY, whereas if it contains a sizeable proportion of particles between 1/256 and $\frac{1}{16}$ millimetre in diameter it is called silt. Consolidated mud, clay and silt are known as mudstone, claystone and siltstone respectively. They are all member of the argillaceous group of sedimentary rocks. Shale is the name given to the variety of mudstone which is fissile, that is capable of being split into very thin sheets parallel to the closely spaced bedding or lamination planes in the rock. Muds are generally of recent origin, clays are abundant mainly in rock sequences younger than the Palaeozoic, while mudstones are common in rock sequences of many ages. The term pelite or pelitic is used by some geologists as a synonym for argillaceous, but it is more commonly reserved for the metamorphic equivalents of argillaceous rocks. Mud rocks are deposited in most sedimentary environments, particularly in bodies of relatively still water.

Mullite. ⟡SILLIMANITE GROUP OF MINERALS.

Muscovite. ⟡MICA.

N

Nacrite. ⇨CLAY MINERALS.

Native element. An element which occurs in nature in an uncombined state. The element may be a metal (e.g. gold), a semi-metal (e.g. arsenic) or a non-metal (e.g. sulphur).

Natron. ⇨SODIUM CARBONATE AND SULPHATE MINERALS.

Natural gas. The term commonly reserved for mixtures of naturally occurring HYDROCARBON and non-hydrocarbon gases often found in subsurface rock formations and commonly associated with crude oil, being in effect the highest or lightest PETROLEUM fractions. Other accumulations result from the alteration of COAL seams to give large quantities of gas and a little tar which migrates into suitable reservoirs or 'fields'. 6000 cubic feet of gas are generally regarded as equivalent to one barrel of oil, and 1967 estimates of world reserves were about 600×10^9 barrels. One cubic foot of natural gas yields about 1032 Btu (British thermal units) of heat.

Natural gas was known in the Middle East in antiquity and the Chinese piped natural gas in the third century AD. The first commercially drilled and exploited gas well was in New York in 1820.

Natural gas is now often transported in a liquefied, refrigerated condition. It is obtained from most OILFIELDS but also from formations in areas such as the North Sea where the gas may be quite separate from OIL POOLS.

Estimates of 'proved' gas reserves are notoriously difficult to make. The US Geological Survey has calculated that somewhere around 30000 to 35000 trillion cubic feet of gas may have been originally in place around the world, of which 19100 trillion cubic feet may eventually be discovered and some 15300 trillion cubic feet recovered. About half the world total natural gas is located in the USSR. Most estimates indicate that the world supply of natural gas will have been used by the end of the century, if the present rate of increase in consumption continues. (⇨ENERGY.)

Natural Glauber's salt. ⇨SODIUM CARBONATE AND SULPHATE MINERALS.

Nepheline. A hexagonal FELDSPATHOID MINERAL $((Na,K)AlSiO_4)$. It occurs as a primary constituent in many alkaline igneous rocks, and it also occurs in some metasomatic rocks (◊METAMORPHISM). Most commercial nepheline is mined from a SYENITE in the Canadian Shield of Ontario. It is used as a substitute for feldspar in the GLASS and ceramics industries.

Neutron logging. A subsurface (downhole) geophysical (radiometric) method (◊GEOPHYSICAL EXPLORATION) involving the use of an artificial neutron source lowered down the well to irradiate the wall rock with fast neutrons. These neutrons are slowed by the environment until they become thermal and are eventually captured. Hydrogen nuclei in water, minerals or hydrocarbons are most efficient in slowing the neutrons. Once the neutrons are captured gamma rays are emitted and these are measured by a scintillation counter in the probe. In essence, therefore, a neutron log is a hydrogen log and this is frequently used in both the oil and water industries to determine POROSITY by measuring the pore volume at 100 per cent saturation.

The neutron log is also widely used to measure moisture changes in the unsaturated zone above the WATER TABLE. Small hand-operated probes are used in this near-surface work.

As with other geophysical logs, the neutron log is most used in correlation of rock strata and aiding the recognition of rock type. In common with the GAMMA LOG, the neutron log can be made in cased holes drilled with fluids such as air and oil. (◊ELECTRICAL LOGGING.)

Nickel (Ni). Most nickel is used in the form of the metal. When alloyed with steel it imparts strength and resistance to corrosion over a wide range of temperature. Nickel steel is widely used in the chemical industry and in the manufacture of aircraft, automobiles and electrical machinery.

The principal ore mineral is pentlandite ((NiFe)S), which is found in the iron–nickel–copper sulphide–platinoid association of basic and ultrabasic igneous rocks. The Sudbury and Thompson Mine deposits (Canada) and the Kambalda deposits (Australia) are of this type. Low-grade, but extensive, residual deposits of nickeliferous LATERITE are also of economic importance. Here the nickel occurs as the silicate – garnierite. In the future, the MANGANESE NODULES of the ocean floors may constitute an important source of nickel. (◊BACTERIAL EXTRACTION OF METALS.)

Major nickel producers are Canada, New Caledonia, the USSR, Cuba and Australia. Total world production (non-Communist) in 1974 was an estimated 1200 million lb.

Niobium. ◊COLUMBIUM.

Nitrate. Most of the cations and anions in ground water can commonly be related to geological factors. However, nitrate in ground water is difficult to relate to such factors as it is such an extremely soluble mineral and is rarely found in rocks. Most nitrate in ground water is derived from nitrogenous organic matter of animal origin, though nitrogenous fertilizers may be a source in ground waters of fissured, UNCONFINED AQUIFERS. The presence of nitrate may thus indicate previous sewage or manurial pollution. In small amounts in water, nitrate is relatively harmless, but if present in amounts in excess of 45 ppm the water will be undesirable for domestic purposes because of the possible toxic effects that it may have on young children. This effect is known as cyanosis: the baby becomes listless, drowsy and his skin takes on a blue colour. Nitrate in drinking water does not cause cyanosis in adults or older children.

Nitrate cannot be removed from water by boiling, but only by demineralizing water or by distilling it. Although pollution may cause local high nitrate values in ground waters of fissured, unconfined aquifers, these values may decrease downdip in the CONFINED AQUIFER owing to the presence of nitrate-reducing bacteria in the waters.

Nitrate deposits. The two commercially important nitrate minerals are nitre or niter (KNO_3) and nitratine ($NaNO_3$). The former is also known as salpetre, and the latter is also known as Chile salpetre or soda-nitre (soda-niter). Both minerals are exceptionally soluble.

The most significant nitrate deposits are situated on the eastern slopes of the Coast Ranges of the Atacama Desert in Chile. They occur in near-surface sands, gravels and clays which are known as caliche. Although the origin of the nitrate component of the Chilean deposits is disputed it is agreed that their crystallization from water requires exceptional aridity; thus they are one category of EVAPORITE DEPOSITS. It has been suggested that the nitrate is leached from guano (bird droppings), that it is fixed by electrical discharges during thunderstorms, that it is fixed by bacteria, or that it is derived from nearby old volcanic rocks.

Nitrate deposits of little or no commercial value are present in some other extremely arid parts of the world.

Until the First World War almost all the world's needs for nitrates were supplied by Chile. At present nitrate minerals contribute less than 10 per cent of the world's supply of nitrates, the remainder being manufactured. Nitrates are used mainly in the chemical industries, especially in the manufacture of FERTILIZERS and explosives. IODINE is a by-product of the processing of the Chilean deposits.

Nitratine. ◊NITRATE DEPOSITS.

Nitre. ◊NITRATE DEPOSITS.

Non-ferrous minerals. A term used by economic geologists for the source minerals of those metals not normally added to iron to make steel alloys. Aluminium, copper, lead, zinc, mercury and magnesium are the main non-ferrous metals. (◊ FERROUS MINERALS.)

Non-metallic rocks and minerals. Whereas knowledge of the categories of materials worked for fuels or metals is relatively widely appreciated, the types of deposits exploited for non-metallic industrial uses are less well known. There is no uniform classification of these deposits which is both scientifically logical and which reflects their many and varied uses. Many non-metallic deposits of a single rock or mineral are put to a variety of uses, principally in the construction, chemical, glass, ceramic, metallurgical, agricultural, paper and food industries. Thus any classification of non-metallic deposits based on their uses will include the same rock or mineral in several categories.

Perhaps the most convenient way of classifying these deposits is based on broad geological associations. The following table gives some idea of a possible classification, and lists many of the rocks and minerals described in detail elsewhere in this book.

Igneous rocks	andesite, basalt, diorite, dolerite, gabbro, granite, peridotite, perlite, pumice, rhyolite, syenite, trachyte
Igneous minerals	cryolite, feldspar, mica, nepheline, vermiculite
Sedimentary rocks	clay, conglomerate, diatomite, dolomite, flint, ganister, gravel, guano, limestone, marl, mud, mudstone, phosphate rock, sand, sandstone, silt, siltstone
Sedimentary minerals	anhydrite, bentonite, borate minerals, gypsum, nitrate minerals, salt, sulphur
Metamorphic rocks	marble, gneiss, quartzite, slate
Metamorphic and metasomatic minerals	andalusite, asbestos, corundum, garnet, graphite, kaolin, kyanite, pyrophyllite, serpentine, sillimanite, talc, vermiculite
Vein minerals	barite, calcite, dumortierite, fluorite, magnesite, topaz, witherite

The commercial value of industrial rocks and minerals is commonly underestimated because some of them do not figure prominently in world trade. The distinction between the industrial rocks and the industrial minerals is as much economic as it is geological. Most industrial rocks occur in deposits of large bulk but they are of relatively low monetary or unit value. They possess a high PLACE VALUE because of the cost of transporting large quantities of material of low unit value. By contrast many of the industrial minerals are rare, and thus worked in relatively small quantities. They occur in deposits of high unit value but of low place value.

Nuclear fuels.   ◊FUEL; URANIUM.

O

Observation well. A well drilled into an aquifer for the purpose of making observations such as water level or pressure recordings. Observation wells are commonly used during the PUMPING TEST of a ground water source or they may be used for long-term observations of natural ground water level fluctuations, rest water levels in areas of aquifers removed from any pumping influence.

Observation wells should be just large enough to allow accurate and rapid measurement of the water levels. During a pumping test on a PRODUCTION WELL the observation well water levels are carefully noted in order to produce a precise picture of the response of water levels in the aquifer to abstraction of water from the pumped well. This monitoring allows analysis of the pumping test data to be made which, in turn, enables long-term predictions of the aquifer yield to be computed.

The observation wells, prior to a pumping test, are normally installed to about the same depth as the production (pumped) well and frequently at 30–100 metres from the pumped well. Observation wells may, however, be terminated in strata above or below the one tapped by the production well to see if there is any HYDRAULIC CONTINUITY between the rock formations.

Ochre. ⟡MINERAL PIGMENTS.

Oil. ⟡PETROLEUM.

Oilfield. The area beneath which a pool or pools of natural PETROLEUM is preserved is an oilfield (Figure 32). It consists primarily of a geological structure which has trapped the oil in a permeable formation underlying an impermeable formation. Oilfields may be a few acres in area or hundreds of square miles. Usually several fields occur in relatively close proximity to one another such as in the North Sea, Iran and the Arabian Gulf. These are all areas of warped or folded sedimentary rocks lying at depths of as much as 9000 metres. OIL POOLS deeper than this are not commonly profitable to tap.

Most oilfields also contain some NATURAL GAS above the oil and BRINE below it. (⟡CONSERVATION; RESERVOIR ROCK.)

Oil pool. An underground accumulation of crude oil trapped in a RESERVOIR ROCK. Many pools may be at depths of less than 1000 metres and the deepest may be at 9000 metres. Coalescing or grouped pools constitute an OILFIELD. The shapes of oil pools are controlled by the structure of the reservoir rock and the nature of its impermeable

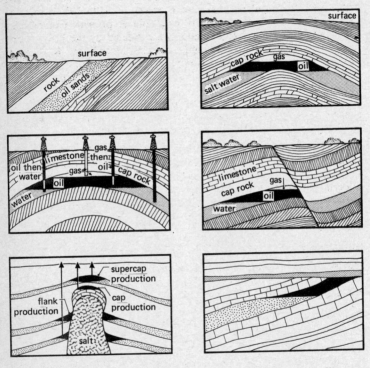

Figure 32. Oilfields occur where geological conditions trap the oil produced from petroleum source rocks. The commonest geological conditions for this are shown.

cover. In plan they are usually round or elongate. Most reservoir rocks are geologically rather young, with the highest ratio of oil pools to volume of sedimentary rock being found in strata of Pliocene (two and a half million years) age.

Oil shale. A fine-grained sedimentary rock containing much original (i.e. in place) bituminous organic matter, incorporated when the sediment was deposited. Oil shale formations are known from Silurian and later rock systems; the richest may yield four barrels of PETRO-LEUM per tonne of rock but the average yield is about 1 barrel per tonne.

The world's largest reserves are the (Eocene) Green River shales of Wyoming, Colorado and Utah, USA. They were deposited in three large shallow lakes. Green River shale may yield 1–1·5 barrels of oil per tonne. The total reserves are in excess of 850×10^9 barrels, of which 240×10^9 are commercially workable at present. Commercial exploitation on a huge scale is now contemplated by the US government. Previous scattered attempts by private industry have until recently been disappointingly costly. A major drawback is that the waste production, mostly dust, is about 150 per cent of the original volume of the shale.

The second largest world deposit of oil shale is in Brazil. Other oil shales occur in the Midland Valley of Scotland (Carboniferous), southern England (Jurassic) and Germany (Jurassic).

Oil trap. The impounding of a 'pool' or 'field' of oil or gas within a RESERVOIR ROCK by the presence of impermeable formations on all sides and above. Traps may be produced by the layering of sedimentary materials or by earth movements giving rise to arched or domed strata, faults, salt domes, etc. (Figures 32 and 34). Most oil traps lie between depths of 1000 and 2500 metres but traps as deep as 4000 metres may yield oil.

A. I. Levorsen, *Geology of Petroleum*, W. H. Freeman, 2nd edn, 1967.

Olivine. The olivine group of silicate minerals form a solid-solution series in which forsterite (Mg_2SiO_4) is one end member and fayalite (Fe_2SiO_4) is the other. Commercially the only olivines of interest are those rich in forsterite. Olivine is a common primary constituent in many basic and ultrabasic igneous rocks such as basalt, gabbro and peridotite. Forsteritic olivine is also a characteristic mineral of some metamorphosed siliceous dolomites.

Olivine is worked only from those rocks in which it is concentrated. It requires careful sorting from associated minerals before it is of value for industrial purposes. MAGNESIUM sulphate has been manufactured from olivine and SERPENTINE. Forsteritic olivine is also used as a REFRACTORY material in some bricks, and in some FOUNDRY SANDS. It has also been applied directly to the soil as a FERTILIZER, and used as a raw material for the manufacture of magnesium sulphate.

Oolites. ◁LIMESTONE.

OPEC (Organization of Petroleum Exporting Countries). ◁FOSSIL FUELS.

Opencast (strip) mining. Removal of a bed of COAL, ORE (1) or other material after stripping away overlying formations. This type of operation is usually carried out in very large pits or excavations of shallow depth for flat-lying or unfaulted deposits. It is now considered profitable to mine opencast for coal to a depth of 60 metres in Europe and North America. The need to move large quantities of earth has led to the development of some of the world's biggest machines, including the 'walking' draglines. In most industrial countries it is now obligatory to return the OVERBURDEN and restore the land surface. Great environmental damage was caused in the past by failure to restore the ground. Water resources have been seriously polluted or disturbed by opencast mining. ($\lozenge$MINING.)

Opencast Mining/Quarrying and Alluvial Mining, Proc. of Symposium, November 1964, Institution of Mining and Metallurgy, 1965.
J. Sinclair, *Quarrying, Opencast and Alluvial Mining*, Elsevier, 1969.

Ore (1). A concentration of a MINERAL or aggregate of minerals (usually metalliferous) which is being worked by a miner, mining company or government.

In most cases, the mineral must be worked at a profit and this is controlled as much by economic and political as by geological considerations. Depending on the price of the metal, a specific mineral deposit may be an ore one week but not the next. Major producers or world powers may manipulate the price of a metal by stockpiling, and then subsequently releasing quantities of the source mineral on the open market. Future technological improvements in extraction techniques may permit the exploitation of mineral deposits currently regarded as having too low a metal content to constitute an ore. On the other hand, the discovery of substitutes for a metal, with a resulting decrease in demand, may make present-day ores unprofitable in the future. In times of war, a government may be compelled to subsidize the working of a nation's own lean deposits of a mineral; in peace, the same deposit may be left in the ground because it is cheaper to import. With these variables and many others, it is virtually impossible to estimate the long-term ore reserves of the earth.

The metal content of the various ores differs widely; currently an iron ore contains 20–30 per cent iron whereas an ore containing only 0·5 per cent copper may be worked profitably. (Examples of the approximate metal content of the common ores compared with the average crustal abundance of the metal are given in the individual element sections.) The present demand for some rare metals, such as GALLIUM or GERMANIUM, is so low that they can be only profitably recovered as a by-product during the processing of an ore for some other metal.

Ore (2). The term ore is often loosely used to describe any noticeable concentration of a metalliferous mineral, whether or not it is of economic value.

Ore (3).   Any mineral which is opaque to transmitted light when examined under the polarizing microscope. (◊ORE MICROSCOPY.)

Ore dressing.   The processing of raw material won by mining into a marketable form; the crushing, concentration and separation of the ORE mineral(s) from the GANGUE. May involve hand-picking, gravity concentration, magnetic separation, dense-media separation or chemical separation.

Ore microscopy.   The study of opaque ore minerals (◊ORE(3)) under the microscope using reflected, as opposed to transmitted, light. Polarized light, etching, microchemical and microhardness tests may also be used.

Orogenic belts, zones.   ◊THE EARTH; FOLD.

Orthoclase.   ◊FELDSPAR.

Orthomagmatic stage.   ◊MAGMA.

Osmium.   ◊PLATINUM-GROUP METALS.

Outcrop.   The area over which a rock FORMATION emerges at the earth's surface. Many outcrops are heavily covered in soil, or other superficial (i.e. unconsolidated) deposits or by vegetation, industrial and urban sprawl or roads. Outcrop mining refers to mining a rock body over its exposed surface as distinct from the mining of concealed, underground formations.

Overburden (1).   Worthless consolidated or unconsolidated material which overlies a deposit of useful material. Used especially in OPEN-CAST (STRIP) MINING.

Overburden (2).   Any loose unconsolidated material which overlies solid bedrock. The terms 'drift', 'mantle material' and 'surface material' may be roughly synonymous.

Oxidized enrichment.   ◊SECONDARY ENRICHMENT.

P

Palladium. ◇PLATINUM-GROUP MINERALS.

Palygorskite. ◇CLAY MINERALS.

Peat. ◇COAL.

Pegmatite. A coarse-grained crystalline rock, commonly composed of quartz (◇SILICA) and FELDSPAR and the result of late-stage crystallization of minerals from a MAGMA that has already given rise to a GRANITE or allied rock.

Pegmatitic stage. ◇MAGMA.

Pelite, pelitic. ◇MUD AND MUDSTONE.

Pellicular water. Water which adheres to soil or rock particles by HYGROSCOPIC and capillary forces. It is non-moving water commonly associated with the INTERMEDIATE ZONE of the AERATION ZONE (Figure 7). (◇SPECIFIC RETENTION.)

Pentlandite. ◇NICKEL.

Perched aquifer. A special case of an UNCONFINED AQUIFER which occurs when a ground water body is separated from the main ZONE OF SATURATION by a rock unit of relatively low permeability and limited areal extent within the AERATION ZONE (Figure 33). Clay lenses in shallow sedimentary aquifers commonly have perched water bodies overlying them. Wells tapping perched aquifers yield only temporary or small quantities of water.

Percolation. The movement of ground water through saturated void and interstitial space (◇INTERSTICES) of an aquifer in the direction of the HYDRAULIC GRADIENT. Percolation is, therefore, movement which takes place below the WATER TABLE (Figure 7) essentially in a lateral, downdip direction. It contrasts with INFILTRATION which takes place above the water table and in a vertical, downward direction from the ground surface.

Perennial streams. Streams which flow above surface at all times throughout the year. Even during the most severe drought periods, the

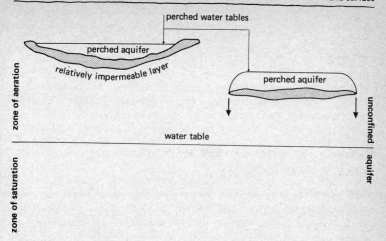

Figure 33. Perched aquifers with the zone of aeration of an unconfined aquifer.

WATER TABLE of the ground water never drops below the stream bed and therefore maintains a continuous supply.

Peridotite. A coarse-grained, plutonic igneous rock of ultrabasic composition. It characteristically contains the mineral OLIVINE, and may in addition contain augite (◊PYROXENE). Feldspar is usually absent in peridotites. Economic geologists are interested in peridotites because ASBESTOS or TALC may be associated with them, and they may be a source of olivine.

Perlite. ◊EXPANDABLE ROCKS AND MINERALS.

Permeability. ◊HYDRAULIC CONDUCTIVITY.

Permeameter. A device to measure the HYDRAULIC CONDUCTIVITY (permeability) of rock or soil samples. It is a laboratory test to give data on the intergranular hydraulic conductivity of the sample. The test is made by measuring the discharge or time rate of change in head for the flow of measured quantities of water through a known area and volume of a soil or rock sample. For relatively permeable materials, such as gravel and sand, a constant-head permeameter is used where the discharge rate from the permeameter is measured. However, for relatively impermeable materials, such as silt and clay, a variable-head permeameter is used where the rate of change of head, rather than the outflow, is measured. Permeability can also be measured with gas

143

permeameters, rather than water, and this technique is practised extensively in the oil industry.

Petroleum. The general term given to crude oil and NATURAL GAS, a complex, naturally occurring, liquid mixture. These materials are found in pores and cavities in rocks into which they have migrated from sources elsewhere. They are HYDROCARBONS which are chiefly, though not exclusively, associated with marine sedimentary rocks. It is from the organic (plant and animal) matter incorporated within these sedimentary rocks that petroleum is derived. Limestones and shales appear to be the most productive source beds but petroleum accumulates in porous rocks (sandstones or fractured strata) in most cases above the source formations.

Crude oil contains about 80 per cent carbon and 11 per cent hydrogen with 1–2 per cent of sulphur, oxygen and nitrogen. Natural gas may have only 65 per cent carbon and up to 25 per cent hydrogen with mere traces of other elements. In most OILFIELDS oil and gas occur together. Usually the gas lies above a layer of crude oil which in turn may overlie salt water.

Petroleum is geographically widely distributed in the major sedimentary basins of the continents (including the CONTINENTAL SHELVES) and is found in rocks of all ages. Most of the large oilfields are in Mesozoic and Cainozoic rocks, but about half of the USSR production is from Palaeozoic strata. Some 60 per cent of known world oil reserves are in Tertiary rocks, about 25 per cent in Mesozoic and 15 per cent in Palaeozoic strata. The two major petroliferous sediment-filled basins are those of the Gulf of Mexico and the lands of the Middle East between the Mediterranean and Iran. They may contain between them 80 per cent of the world's oil.

Oil, ASPHALT and natural gas were known in the Mediterranean and Persian regions several hundred years BC. 'Mineral oil' has been used for various purposes throughout history and was sought by the Chinese many centuries ago. It was, however, not until 1859 that crude oil was obtained through a borehole from a subsurface OIL POOL. Only then did it become obtainable in sufficient quantity to influence the course of industry. (◊CONSERVATION; ENERGY.)

E. N. Tiratsoo, *Oil Fields of the World*, Scientific Press, 1973.
A. I. Levorsen, *Geology of Petroleum*, W. H. Freeman, 2nd edn, 1967.

pH (hydrogen ion concentration). The relative concentration of hydrogen ions in water indicates whether the water will act like a weak acid, or if it will perform as an alkaline solution. When the amount of hydrogen ions is excessive in relation to the other ions in the solution, an acid reaction results. These waters tend to attack metals.

Neutral water has a pH of 7, which indicates an equal number of hydrogen and hydroxyl ions. When the water has a pH value greater than 7 the water is alkaline in type and when the pH is below 7 the water is acidic in type. ($\Diamond$ ACIDITY; ALKALINITY.)

When material goes into solution in water the pH is commonly changed. Most ground waters have pH values between 5 and 8. Very high pH values above 8·5 are usually associated with sodium carbonate–bicarbonate waters. Very low pH values below 4 are associated with waters containing free acids.

Phlogopite. $\Diamond$ MICA.

Phosphate deposits. Phosphates, which are essential for life, are mainly used in FERTILIZERS. The two principal types of phosphate deposit are rock phosphates and the mineral apatite ((CaF)Ca$_4$(PO$_4$)$_3$).

The rock phosphates include: phosphatic limestone, a rock in which some of the calcite has been replaced by calcium phosphates; phosphorite, a rock consisting of cryptocrystalline calcium phosphate in the form of collophone (Ca$_3$P$_2$O$_8$.H$_2$O) and calcite; and guano, a deposit formed from an accumulation of bird droppings. Guano occurs on some oceanic islands and on some desert coasts, as, for example, in Peru and Chile. Some phosphatic limestones form as a result of phosphoric acid, derived by downward leaching from overlying guano deposits, reacting with calcite. Phosphorite may be deposited as primary layers beneath the sea, especially, it is thought, in a reducing environment, as for example on the floor of the Black Sea. It is also thought that the presence of strong, cold, submarine currents encourages phosphate deposition. Some phosphate deposits are of residual or alluvial origin, the phosphate being derived from pre-existing rocks. Important rock phosphate deposits occur in Tertiary rocks in Florida, and in Permian rocks in some Rocky Mountain States. The phosphate deposits of Morocco and adjacent territories occur in bedded sequences.

The other principal source of phosphates is apatite, a mineral which is concentrated in commercial quantities in some PEGMATITES, as, for example, in Quebec. Apatite is also associated with GABBROS in Norway. Near Murmansk in Russia an apatite-nepheline igneous rock is worked for apatite. Overwhelmingly the greatest use of phosphate deposits is the production of fertilizers. Lesser quantities of phosphates are used in water softeners, baking powder, and in the manufacture of detergents, REFRACTORY bricks, ferrophosphorous and plastics. Some of the interest in phosphates centres on the small percentage of URANIUM oxide which occurs in association with them. Similarly VANADIUM occurs in association with phosphate deposits in Idaho. Some phosphate is obtained as a by-product from the slag of basic steel furnaces. Production in 1974 in tonnes of phosphates content was: USA, 41 500 000; USSR, 22 540 000; Morocco, 19 326 000.

Phosphoric limestone. $\Diamond$ PHOSPHATE DEPOSITS.

145

Phosphorite. ▷PHOSPHATE DEPOSITS.

Phreatic surface. ▷WATER TABLE.

Phreatic water. ▷GROUND WATER.

Phreatophytes. Desert plants with deeply penetrating roots which reach the WATER TABLE. They are localized along stream courses and are also found in moist environments, but the ecological classification is not distinct. Certain phreatophytes have a low tolerance for salt and are thus valuable guides to potable water in arid and semi-arid regions. Willow, ash, cottonwood and alder are useful in this regard, normally growing where the water table is less than 10 metres below ground surface.

Piezometric surface. When a well is drilled through the upper AQUI-CLUDE into a CONFINED (ARTESIAN) AQUIFER, ground water will rise in that well to a level above the top of the aquifer (Figure 9). The water level in the well represents the hydrostatic or artesian pressure level in the aquifer. The piezometric surface of a confined aquifer is an imaginary surface coinciding with the hydrostatic pressure level of water in the aquifer. The ground water level in a well penetrating such an aquifer gives the elevation of the piezometric surface at that point. If the piezometric surface is below ground level, the well tapping the confined aquifer will be a non-flowing ARTESIAN WELL. Should the piezometric surface lie above ground surface a flowing or overflowing artesian well results.

The piezometric surface is analogous to the real water surface, the WATER TABLE, in an UNCONFINED AQUIFER. Contour maps and profiles of the piezometric surface can be prepared from well data similar to those for the water table in an unconfined aquifer. Piezometers are pressure-reading and measuring instruments for water levels in either unconfined or confined aquifers.

Pinite. ▷SILLIMANITE GROUP OF MINERALS.

Pipe and tile clays. CLAYS used for making drain and sewer pipes, tiles and architectural terracotta are generally similar to those used for making bricks (▷BRICK CLAYS), but they are processed more carefully. They should not contain an excess of iron oxides, and the iron oxide content may need to be controlled to maintain a constant colour in the product.

Pitch. ▷ASPHALT.

Pitchblende. ▷URANINITE.

Place value. Deposits of high place value are those which are widespread and occur in considerable bulk, but which are of low MONETARY

OR UNIT VALUE. They include most of the rocks used in the construction industry. They are generally processed simply and cheaply, and transported for only short distances. They are rarely exported. Although they are of low unit value their total value commonly exceeds that of many deposits of high unit value.

Placer deposit. A sedimentary deposit, usually of gravel or sand, containing a natural concentration of a valuable mineral. Placer minerals are characterized by resistance to chemical and mechanical weathering and a higher density than the common rock-forming minerals. Concentration is effected by moving water, usually streams but sometimes marine currents, and occurs wherever natural barriers trap heavier or coarser particles and allow the lighter or finer particles to be carried away.

Gold, platinum, diamonds and tin (cassiterite) form important placer deposits. Geologically ancient placer deposits are known. According to some authorities, the Precambrian gold-bearing CONGLOMERATES (banket) of the Witwatersrand are of such an origin.

Platinum-group metals. Besides platinum (Pt), the group includes palladium (Pd), iridium (Ir), osmium (Os), rhodium (Rh) and ruthenium (Ru). The metals are characterized by their resistance to corrosion and oxidation, good electrical conductivity and good catalytic activity. They are used especially in petroleum refining, in the production of chemicals, and in electrical components. Much interest in recent years has been shown in the possible use of platinum as a catalyst to remove pollutants from motor car exhaust fumes.

The metals of the platinoid group commonly occur in the native state (often alloyed with one another), and as arsenides and sulphides. The most important commercial deposits are associated with magmatic nickel–copper sulphide deposits occurring in ultrabasic and basic rocks.

World production for platinum in 1974 was an estimated 2·8 million troy ounces. The USSR, South Africa and Canada were the only significant producers.

Plumbago. ⬦GRAPHITE.

Plunge. ⬦FOLD.

Plutonic rocks. ⬦IGNEOUS ROCKS.

Pneumatolytic stage. ⬦MAGMA.

Pollucite. ⬦CESIUM.

Polyhalite. ⬦POTASH DEPOSITS.

Porcelain clays. Porcelain is manufactured from clay, feldspar and quartz. Some high-quality British porcelain is made from china stone

($\Diamond$CHINA CLAY). The two types of clay most commonly used are BALL CLAY and china clay.

Pores. Small VOID spaces in rocks or in unconsolidated materials such as aggregates of soil particles; synonymous with INTERSTICES. These are important in HYDROGEOLOGY as interconnected pores act as conduits for subsurface waters. A porous medium is one that contains interdispersed void or pore space, e.g. an aquifer. The pore space is the space occupied by voids, containing gases or liquids, in a rock or soil sample. The pressure of water in pores of a saturated porous medium is referred to as the pore pressure. ($\Diamond$POROSITY.)

Porosity. A porous medium possesses a solid matrix or skeleton, an assembly of solid mineral grains separated and surrounded by VOIDS, PORES or INTERSTICES which may be filled with water, gases or organic matter. The porosity of a rock or soil is a measure of the contained interstices or pores. It is expressed as the percentage of void space to the total volume of the mass.

In HYDROGEOLOGY porosity is important in assessing the volume of water in storage related to the total volume of the aquifer. Porosities in granular sedimentary aquifers depend upon the shape and arrangement of individual grains, the variation in grain size, and the degree of cementation and compaction. In solid, bedrock aquifers the removal of mineral matter by solution and degree of rock fracture are also important. Porosities range from near zero to in excess of 50 per cent depending upon these factors and the type of material. Total porosity comprises specific yield plus SPECIFIC RETENTION of a porous medium. EFFECTIVE POROSITY is synonymous with specific yield.

Porphyry copper. A term originally used to describe the ore of Bingham Canyon, Utah, where hydrothermal COPPER minerals are disseminated in porphyry – an igneous rock characterized by large crystals of quartz or feldspar set in a matrix of finer-grained crystals. Now used to describe any large amount of disseminated ore, of hydrothermal origin, irrespective of the nature of the host rock. The ore is low grade (it may contain less than 0·5 per cent copper) but the copper is evenly distributed throughout an immense volume of rock. Thus the porphyry copper of Bougainville, Papua New Guinea, has an estimated reserve of 900 million tonnes of 0·48 per cent ore.

Porphyry coppers account for about half of the world's production and are usually mined in large open-pit excavations or massive underground mines. Mining on such a scale presents many problems of environmental pollution.

Portland cement. $\Diamond$CEMENT.

Potash deposits. Potash is the name given to potassium oxide (K_2O), a compound which does not exist in nature. Although many minerals contain some potassium the principal commercial sources of potash are four highly soluble minerals in some EVAPORITE DEPOSITS. They are sylvite (KCl); langbeinite ($K_2Mg_2(SO_4)_3$); carnallite ($KMgCl_3$. $6H_2O$); and kainite ($MgSO_4$. KCl. $3H_2O$). Where sylvite is inter-grown with halite ($\Diamond$ SALT) it is known as sylvinite. A related and more abundant potassic mineral is polyhalite ($2CaSO_4$. $MgSO_4$. K_2SO_4. $2H_2O$), but it is rarely worked. The potash minerals are generally thought to have crystallized during the final stages of evaporation of ultrasaline bodies of seawater.

Economically significant potash minerals generally occur in associa-tion with other evaporite minerals. They are commonly worked by underground mining, or by pumping water underground and extracting the resulting potassic brine. The extreme solubility of the minerals means that they rarely crop out at the surface.

The best-known potash deposits are those of Stassfurt in north-central Germany. The Stassfurt deposit occurs in rocks of Permian age, and in association with anhydrite and salt. Similar Permian potash deposits, with an estimated reserve of about 400 million tonnes, occur 3000 feet underground in North Yorkshire, and are mined at Boulby. The potash deposits within the upper part of the Rhine Rift Valley in the province of Alsace in France are in rocks of Oligocene age.

In the USA important potash deposits are worked from rocks of Permian age near Carlsbad in New Mexico. Deep deposits of high-grade potash salts, together with rock salt, anhydrite and dolomite, occur in rocks of Devonian age beneath the plains of Saskatchewan. Potassium chloride is obtained from BITTERNS remaining after the processing of the brines of several salt lakes in the western USA. It is also obtained from the brines of the Dead Sea.

Some non-evaporitic minerals, including alunite, ALUM, LEUCITE and GLAUCONITE, are potential sources of potash.

The principal use of potash that has been derived from minerals is in the manufacture of FERTILIZERS, lesser amounts being employed in various chemical industries and in the manufacture of soap and deter-gents, dyes, drugs, explosives, textiles and ceramic products. Production in 1974 in tonnes of potassium oxide content was: USSR, 6500000; Canada, 5496000; West Germany, 2865000. ($\Diamond$ GLASS.)

Potential logging. $\Diamond$ ELECTRICAL LOGGING.

Pottery and stoneware clays.   The requirements of a CLAY that is to be used in the manufacture of pottery are less rigorous than those for the manufacture of higher-grade ceramic products. Nevertheless semi-REFRACTORY CLAYS are required. They should be plastic ('fat'), should not shrink excessively, and when fired they should be strong.

In the USA, suitable clays are widespread; many are FIRECLAYS of Upper Carboniferous age from the central States, or of Cretaceous or Tertiary age elsewhere.

Stoneware clays are commonly silica-rich clays containing some free quartz. In Britain they are worked from beds beneath the BALL CLAYS in the Tertiary sequence of the Bovey Tracey Basin in South Devon.

ppm (parts per million). A method of recording small quantities. For example, the average crustal abundance of tantalum may be expressed as either 0·0002 per cent (parts per hundred), or, more briefly, as 2 ppm. Parts per million are equivalent to grammes per tonne; thus a tonne of crustal rock contains on average 2 grammes of tantalum.

Precious metals. GOLD, SILVER and the PLATINUM-GROUP METALS. Mineral statistics for the precious metals are usually quoted in troy ounces. One troy ounce = 31·1035 grammes.

Precipitation. All forms of water, in liquid or solid form, deposited on the earth's surface and derived from atmospheric vapour (mist, rain, hail, sleet and snow). Condensation on solids and water surfaces as dew and frost is sometimes considered a form of precipitation.

Precipitation on land areas is the source of essentially all freshwater supplies. It replenishes the quantity that is taken from lakes, streams and wells for man's use.

Precipitation is a major factor in the HYDROLOGICAL CYCLE and is intimately linked to the EVAPORATION process. Total precipitation from the atmosphere is equal to the evaporation into it. It is estimated that some 400 000 cubic kilometres of water fall back to the earth each year. Of this amount the land surfaces of the earth receive about 100 000 cubic kilometres of freshwater yearly.

Primary energy sources. Energy arrives at the surface of the earth from two principal directions, down and up. Coming down is solar radiation, sunshine, heat, varying with the season of the year, time of day and weather. There is also the gravitational pull of the sun, moon and planets which produces the tides. Coming up is the earth's internal heat. It is, however, solar energy which is the driving force behind atmospheric circulation, ocean currents and biological activity which man may harness as sources of power.

Production well. A well from which ground water is pumped for industrial, agricultural or domestic purposes (abstraction well or pumping well).

Production wells may be of any depth or diameter depending on geological conditions at the well site. The depth of the production well may be known after the lowest rest water level has been found in the area and consideration has been given to the projected pumping water level (Figure 18). It is often difficult to offer guidance on the diameter of the production well, since this depends upon local geological factors, the projected well yield and, hence, the diameter of the submersible pump. Too small a diameter will limit ground water supplies to the production well and produce high entrance velocities around the well. Too large a diameter may be economically unsound because of the high cost of casing. Generally a 45-centimetre diameter production well is the optimum size. However, production wells vary in diameter from 15 to 110 centimetres; many large public water supply pumping stations use 90-centimetre diameter wells.

Psilomelane. A complex hydrated manganese oxide containing varying amounts of BARIUM. It is, with CRYPTOMELANE, a common constituent of sedimentary and residual manganese deposits. Psilomelane is also found in exhalative (hot spring) deposits.

Pumice. Highly porous, cellular, natural volcanic glass. It is formed when gases escaping from a rapidly cooling lava leave behind many bubbles and air holes. Most pumice is of acid composition. (◊IGNEOUS ROCKS.)

A minor use of pumice is as an ABRASIVE, especially in scouring stones, but its main value is as a natural lightweight AGGREGATE. (◊MINERAL FILLERS.)

Pumping tests. The most reliable method for estimating permeability and other fundamental hydrological parameters of aquifers is by means of test pumping wells. A pumping test will give information on the SAFE YIELD of an aquifer, the effect of ground water abstraction from the PRODUCTION WELL on the local ground water conditions, and the effect that the abstraction will have on existing ground water producers in the area.

The pumping test is thus designed to determine some or all of the following: the yield characteristics and potential of a well and the prediction of its operating conditions, the hydrogeological properties of contributory aquifers, the effect of present and future abstraction from the well and the safe yield of the production well and the aquifer.

A pumping test programme is divisible into three stages – before, during and after the period of discharge from the well. Before the test a hydrogeological study of the contributory aquifers is undertaken to determine their distribution both at OUTCROP and at depth. Changes in aquifer (rock type) are noted as this affects the permeability of the media. Also the overall conditions of ground water flow are assessed and related to UNCONFINED or CONFINED AQUIFER conditions.

The analysis of pumping test data depends essentially upon fitting it to the theoretical model considered most likely to correspond to the actual hydrogeological conditions. The four most commonly encountered variables are the rate of abstraction in the production well, the

DRAWDOWN of the rest water level in surrounding observation wells, the distance to the point of observation and the time since pumping commenced. These parameters can be varied to different degrees experimentally during a pumping test.

The techniques of pumping test procedure and observations are closely associated with measurements of these four parameters.

Pumping well. ⟡ PRODUCTION WELL.

Pumps. Upon the completion of a PRODUCTION WELL some form of pump must be installed to lift the ground water and deliver it to the point of use. A pump develops no power of its own. Its purpose is simply to transfer energy from a source of power to move a fluid. To satisfy the demand for water this resource has to be raised from wells, lakes and streams. The development of pumps and pumping techniques has stemmed from this need for water.

A variety of pumps can be used for shallow wells and discharge can range from a few to about 6000 gallons per hour, depending upon the pump type and size of intake and discharge pipes. A so-called shallow well pump refers to a pump that is located above the well and takes water from the well by suction lift. Suction lifts should not exceed 6–8 metres for efficient and continuous operation.

The equipment involved in a so-called deep well pump is installed within the well CASING and usually with the pump inlet submerged below the pumping water level. The deep well pump must be used for any well, regardless of depth, where the pumping level is below the limit of suction lift. In large deep wells yielding flows in excess of 30000 gallons per hour, the deep well turbine pump has been most widely adopted.

Pyrite. IRON sulphide (FeS_2); a very common and widespread sulphide. It is pale brass yellow in colour with a greenish or brownish-black streak, and is brittle with a metallic lustre. Hardness is 6–6·5.

Pyrite is usually mined for the gold or copper associated with it but is also used to furnish SULPHUR for the manufacture of sulphuric acid.

Pyroclastic rocks. Accumulations of fragmentary volcanic ejectamenta which have been explosively erupted from a volcano, or fill a volcanic vent. Many pyroclastic rocks are layered in a manner similar to SEDIMENTARY ROCKS. Ash is fine-grained, generally poorly consolidated pyroclastic material. Compacted ash and some coarser-grained pyroclastic material are often called tuff. Agglomerate is pyroclastic material containing large particles. An ignimbrite is an ash flow deposit. Many ignimbrites consist of shards of volcanic glass which have been welded together as a consequence of their high temperature.

Most pyroclastic rocks are of little value except as CRUSHED STONE. Some cleaved tuffs, which are now SLATES, make attractive facing or cladding material. The Lakeland green slates of England are of this type. ($\Diamond$DIMENSION STONE.)

Pyrolusite. The most common manganese ore (MnO_2), containing 63 per cent manganese. It is a soft (hardness 1–2), black, metallic mineral. Pyrolusite is commonly found as a secondary mineral; it may be a constituent of LATERITES.

Pyrophyllite. A soft, pale-coloured mineral ($H_2Al_2(SiO_3)_4$) similar to TALC, found in pod- or lens-like masses in some metamorphic rocks which have resulted from the metasomatic alteration ($\Diamond$METAMORPH-ISM) of acid volcanic and PYROCLASTIC ROCKS. It is worked in North Carolina, California and Newfoundland, and is found elsewhere in the world.

Although pyrophyllite was originally employed as a firestone or hearthstone, many of its present uses are similar to those of talc. It is employed as a REFRACTORY material, and as a filler ($\Diamond$MINERAL FILLERS) in many products including paint, wall boards and insecticides.

Pyroxene. Most of the minerals belonging to this group are anhydrous ferromagnesian silicates. Many pyroxenes are dark-coloured. Their characteristic occurrence, as the mineral augite, is in basic and ultra-basic igneous rocks. Some pyroxenes such as diopside ($CaMg(SiO_3)_2$) occur in metamorphosed impure limestones. Spodumene ($LiAl(Si_2O_3)_2$) is a source of LITHIUM.

Q

Quartz. ▷SILICA.

R

Radium (Ra). The most commonly known radioactive element in the radioactive family. When pure it is a white, shiny metal and several isotopes are known. It disintegrates by the emission of alpha particles. In nature it is found in association with URANIUM in pitchblende and other minerals. It is also present in many natural waters and in marine sediments. Even the richest ores contain only about 0·25 grammes of radium per tonne. The element is produced commercially principally in the USA, Belgium, Canada and the USSR.

Radium is used for its radiation, principally in medicine but also to a minor extent in industrial radiography. At one time radium was used in the luminous paint for the figures on the faces of watches and clocks.

Ranney well. ◊COLLECTOR WELL.

Rare earth elements. The fifteen elements from atomic numbers 57 to 71, most with unfamiliar names:

lanthanum (La)	europium (Eu)	erbium (Er)
cerium (Ce)	gadolinium (Gd)	thulium (Th)
praesodymium (Pr)	terbium (Tb)	ytterbium (Yb)
neodymium (Nd)	dysprosium (Dy)	lutetium (Lu)
samarium (Sm)	holmium (Ho)	

Promethium (atomic no. 61) is a man-made fission product of URANIUM. Although originally considered scarce, in terms of crustal abundance, the rare earths are actually more abundant than many more familiar elements. Thulium, the scarcest rare earth element, is more abundant (7 ppm) than gold, silver and platinum combined.

The principal use of the compounds of rare earths is as catalysts in petroleum refining, in the manufacture of special glasses and various electronic applications. 'Misch-metal' is an alloy of the rare earth metals and is the chief constituent of cigarette lighter flints. COBALT–rare earth alloys are important in the field of permanent magnets.

The commonest commercial source of the rare earths is monazite (a phosphate of thorium and the rare earths), which occurs as PLACER DEPOSITS in the USSR, Australia, India, Brazil and Malaysia. Usually the monazite is a by-product in the mining of TITANIUM and zircon

($\lozenge$ ZIRCONIUM). A unique deposit of bastunaesite (a cerium and other rare earth fluocarbonate) occurs at Mountain Pass, California, and produces the bulk of the total of the US production. Total world production (1974) was of the order of 9000 tonnes of rare earth oxides.

Recession curve. Comprises the falling limb of a river or well HYDRO-GRAPH curve. In rivers the recession flow is that flow which continues after rainfall has stopped. A ground water recession curve shows the variation of BASE FLOW with time during periods of little or no rainfall over a river basin. It is a measure of the drainage rate of ground water storage from the basin. Knowledge of the shape of a recession curve enables estimates of stream flow to be made during drought periods.

Recharge area. Principal sources of natural recharge, or additions of water, to aquifers include PRECIPITATION, stream flow, lakes and RESERVOIRS. Natural recharge to aquifers will take place over the OUTCROP area, or recharge area, of the UNCONFINED AQUIFER where the aquifer is in direct influence of the atmosphere. Natural recharge to CONFINED AQUIFERS may take place by leakage of water through contiguous AQUICLUDES or by deep-seated leakage along FAULT lines.

Apart from natural recharge there may be other contributions of water to aquifers involving artificial replenishment of a depleted aquifer by injection of water through recharge wells or excess INFILTRATION of water from the surface. The latter may include ARTIFICIAL RECHARGE from excess irrigation water, seepage from canals or infiltration from recharge basins or pits. Artificial ground water recharge is not as yet practised extensively, but it may provide a major factor in future water resources development.

Recovery method. During a PUMPING TEST on a well a method may be employed whereby both DRAWDOWN and recovery of head of water after pumping has stopped are observed in the same well. Values obtained from analysis of the recovery record serve to check calculations based on the time–drawdown method of the pumping record.

Redox potential. $\lozenge$ Eh.

Refractories. A refractory material is a non-metallic substance capable of physically and chemically withstanding high temperatures and rapid changes in temperature. Many of the rocks and minerals exploited for their refractory properties are treated elsewhere in this book under the headings of BAUXITE, BRICK CLAYS, CEMENT, CERAMIC CLAYS, CHINA CLAY, CHROMITE, DOLOMITE, FIRECLAY, FOUNDRY SAND, GRAPHITE, MAGNESITE, OLIVINE, PYROPHYLLITE, REFRACTORY CLAYS, SILICA, SILLIMANITE GROUP OF MINERALS, TOPAZ and WOLLASTONITE.

Refractory clays. Refractory clays are similar to CERAMIC CLAYS, but after firing they yield materials capable of withstanding very high temperatures (◇REFRACTORIES). Suitable clays generally contain a high proportion of the CLAY MINERAL kaolin, and they may also contain excess SILICA in the form of quartz grains. The refractory properties of suitable clays are commonly proportional to their alumina content, clays containing significant quantities of iron and lime being generally less refractory. Perhaps the most important refractory clay is FIRECLAY, but some BRICK CLAYS are suitable for the manufacture of refractory bricks.

Remote sensing. Detection of natural resources by various devices installed in aircraft and satellites: radar, gamma ray detectors and sensors of infra-red energy are among the instruments used.

This area of exploration has expanded vigorously in the last two decades as a result of industrial efforts to locate resources and as a consequence of the NASA space programme. Minerals, oil and gas, water and major geological features such as earthquake-producing fault zones may be located by these means. Soil characteristics and distribution can be plotted and vegetation assessed. Oceanographic applications are also important and numerous. Many of the early hopes for successful remote sensing have been achieved by the American ERTS satellite (◇LANDSAT).

R. N. Colwell, Remote Sensing of Natural Resources, *Scientific American*, vol. 218 (1968), pp. 54–71.

Renewable fuels. ◇FUEL.

Reserves. That proportion of a resource which can under the prevailing economic or technological conditions be utilized. As the development of machinery for drilling oil wells at sea has taken place, so the extractable reserves of oil available have increased. Potential reserves of oil may exist in areas too deep for drilling at the present time: one day they could be extractable resources.

Reservoir rock. A geological formation which is porous, fractured or otherwise contains cavities which hold PETROLEUM (Figure 32) or water. Thus most coarse-grained and poorly cemented or irregularly cemented sedimentary rocks are potential reservoirs. Sandstones and limestones are common reservoir rocks. The oil has been trapped in the cavities while migrating from some rocks under pressure. Permian or Jurassic sandstones are reservoir rocks in some North Sea oilfields while many Middle Eastern reservoir rocks are Tertiary limestones. (◇OIL TRAP.)

Reservoirs. A term applied to either a recipient for the collection of a small amount of liquid or a structure used as a surface water impoundment. Public water supplies from surface sources include water taken locally from rivers and lakes, and also that stored in impounding reservoirs generally situated at some distance from the area of supply. Thus a town which lies on a large river frequently takes its water from such a source. The reservoir water is filtered and, if necessary, purified (chemically and bacteriologically) before being used.

Reservoirs are thus used directly as a public water supply source or they may be used for flood control purposes. Peak river flows can be reduced by temporarily storing a portion of the surface RUNOFF until after the crest of the flood has passed. There are two types of reservoir storage, controlled and uncontrolled. In controlled storage, gates in the impounding reservoir may regulate the outflow in any desirable manner. In uncontrolled storage there is no regulation of the outflow capacity of the impounding dam. These structures contain overflow spillways, and the only flood benefits obtained from them result from the modifying and delaying effects of the storage above the spillway crest.

Residual drawdown. After pumping is stopped in a PRODUCTION WELL, water levels rise or recover and approach the rest water level observed before pumping started. During such a recovery period the distance that the water level is found to be below the initial rest water level in the well is called the residual drawdown.

Resistivity logging. ▷ELECTRICAL LOGGING.

Resources. In the sense of earth resources, the term encompasses naturally occurring materials which are of importance to man and his activities. Renewable resources are those which can be replaced by fresh growth – as in crops and forests, or water or solar energy. Non-renewable resources are those which we destroy when using them – coal and oil are two such examples.

The study of resources calls for the services of scientists, technologists and engineers, entrepreneurs for resources and financial specialists and economists. Man's demands have changed with the passage of time. The tendency is for progressively more natural resources to be required and developed. In the process some have been destroyed, others depleted to a very low level. The search for new resources for materials and ENERGY is now conducted on land, sea and in the air with the aid of satellites (▷ERTS; LANDSAT; REMOTE SENSING) and, eventually, Skylab data. (▷CONSERVATION.)

W. N. Peach and J. N. Constantin, *Zimmerman's World Resources and Industries*, Harper & Row, 1972.

Retained water. When ground water is pumped from the ZONE OF SATURATION, not all the water is removed from the porous medium. Some water is held in place by molecular and surface-tension forces. This is termed retained water and is water held in place against gravity.

The amount held can be quantitatively evaluated as SPECIFIC RETEN-TION. FIELD CAPACITY, PELLICULAR WATER and retained water are all forms of specific retention. They refer to the same water content but differ by the zone or subzone in which they occur.

Rhenium (Rh). An extremely scarce metallic element (crustal abundance 0·005 ppm) which has a melting point of 3167°C, exceeded only by tungsten. It is used mainly as a constituent of ductile, high-temperature TUNGSTEN and MOLYBDENUM alloys. A potential use is in the production of rhenium–platinum catalysts in petroleum refining. It is produced as a by-product from molybdenite ore which, in turn, is often a by-product of PORPHYRY COPPER ores. No reliable statistics are available but world production (1974) was probably in the order of 15 700 lb.

Rhodium. ◊PLATINUM-GROUP METALS.

Rhyolite. A pale-coloured, fine-grained, volcanic igneous rock of acidic composition. In addition to natural volcanic glass, rhyolites contain alkali feldspar and quartz. In chemical composition they are equivalent to GRANITE. Rhyolites are mainly worked for CRUSHED STONE. (◊GRANULES.)

Riebeckite. ◊ASBESTOS.

Rip-rap. Large blocks of untrimmed stone for use in the construction industry. Rip-rap is, for example, employed in shore protection schemes. Most rocks quarried for DIMENSION STONE are suitable for rip-rap.

River. A natural water course through which RUNOFF reaches an inland body of water or the sea. There often exist tremendous differences between the various characteristics of two rivers draining areas of nearly the same size and located only a short distance apart. Great flow variations in the rivers can also occur from day to day and year to year on the same river. Thus, because of the variability in river discharge, records for a minimum of some twenty years' flow are required to determine the general nature of the river regimen at any point.

Rivers may be divided into three general classes, each having a characteristic type of runoff depending upon the physical characteristics and climatic conditions of the drainage basin. The classes are EPHEMERAL, INTERMITTENT and PERENNIAL STREAMS.

Road metal (roadstone). Roadstones are hard and durable rocks which can be successfully worked and crushed for use in road making. For the preparation of the upper wearing layer of a road it is important that the rocks chosen are capable of resisting polishing. Rocks used as roadstone include granite, dolerite, basalt, felsite, sandstone and quartzite (a variety of sandstone). The low monetary value of road-stones makes local supplies virtually essential.

Rock. To a geologist a rock is an aggregate of solid mineral particles or, less commonly, organic material, which forms a part of the earth's crust. Although soil is generally excluded from the definition it covers both consolidated and unconsolidated material. Thus both sand and sandstone are rocks. Popular usage, and that of engineers, generally restricts the term to hard or consolidated earth material. The three principal classes of rocks are IGNEOUS, SEDIMENTARY and meta-morphic (◊METAMORPHISM).

Rock crystal. ◊SILICA.

Rock phosphates. ◊PHOSPHATE DEPOSITS.

Rock salt. ◊SALT.

Rock wool. ◊MINERAL AND ROCK WOOL.

Roscoelite. ◊VANADIUM.

Rotary drilling. ◊DRILLING METHODS.

Rottenstones. ◊TRIPOLI.

Rubidium (Ru). Although the ninth most abundant metallic element in the earth's crust – it is more abundant than copper, lead or zinc – rubidium is widely dispersed and forms no natural mineral concentrations.

There are no reliable statistics for rubidium usage; small amounts are used in the manufacture of vacuum tubes and photocells, and also as radioactive rubidium in medical work. All the rubidium of commerce is recovered as a by-product of the processing of CESIUM and LITHIUM ores.

Ruby. ◊CORUNDUM.

Runoff. The discharge of water through surface streams of a drainage basin. Total runoff comprises the sum of surface runoff and ground water discharge that reaches the rivers. Surface runoff equals PRECI-PITATION minus surface retention and INFILTRATION. Surface run-off is commonly represented in the form of a HYDROGRAPH which is a time record of stream surface elevation or stream discharge at a given cross-section of the stream (Figure 31). In the case of a stream with a discharge maintained long after precipitation has taken place, the discharge also includes ground water runoff (BASE FLOW).

The runoff portion of the HYDROLOGICAL CYCLE includes the distribution of water and the path followed by water after it precipitates on the land until it reaches stream channels or returns directly to the atmosphere through evapotranspiration (◊TRANSPIRATION).

The runoff coefficient is a dimensionless coefficient to estimate run-off as a certain percentage of storm rainfall.

Ruthenium. ◊PLATINUM-GROUP METALS.

Rutile. Titanium dioxide (TiO_2), a reddish-brown, yellowish or black mineral. Rutile is an important ore of TITANIUM. Commercial deposits occur chiefly as PLACER DEPOSITS.

S

Safe yield. The amount of water which can be withdrawn from a ground water basin annually without producing a deleterious result. Any withdrawal in excess of safe yield is an overdraft. If ground water is regarded as a renewable or replaceable natural resource, then only a certain quantity of water may be withdrawn annually for a ground water body. The maximum quantity of water which can be extracted from a ground water reservoir, yet still maintains that supply unimpaired, depends upon the safe yield.

Some authors define the safe yield of an aquifer as one which produces no permanent fall in ground water levels or no deleterious change in water quality. Others regard safe yield as the balance between abstraction and INFILTRATION. Safe yield is not a fixed quantity, but depends upon the natural conditions occurring in a particular area.

Saline intrusion. In COASTAL AQUIFERS with hydraulic connection (◊HYDRAULIC CONTINUITY) with the sea, fresh and saline waters will coexist in the aquifer with a mutual boundary known as the saline interface (Figure 16). When fresh, potable ground waters are pumped from wells drilled into the coastal aquifer, the hydrodynamic balance between the two types of water will change. The outflow of freshwater towards the sea will be reduced and the WATER TABLE lowered. As a result the saline interface and the saltwater body will move inland, further into the aquifer. This is known as saline intrusion and is a process which can result in the contamination of the aquifer.

The total ground water flow in the aquifer is equal to the natural recharge from INFILTRATION occurring on land. Part of this flow may be intercepted by pumping wells in coastal aquifers, but to prevent saline intrusion some of the flow must be allowed to proceed in a seaward direction to check the advance of the saline interface towards the pumping wells.

Saline lake. A characteristic feature of some basins of internal drainage in hot, arid regions. They are generally richer in chlorides than ALKALI LAKES, but they may also contain significant quantities of sulphates. (◊BRINE.)

162

Salpetre. ⟡NITRATE DEPOSITS.

Salt.   Common salt is sodium chloride (NaCl). The pure, cubic mineral form which is soft, colourless and soluble in water is called halite. Beds rich in halite but containing impurities are known as rock salt.

Sodium chloride accounts for about 78 per cent of the total dissolved solids in seawater; thus not surprisingly most rock salt deposits occur in sequences of EVAPORITE DEPOSITS which have resulted from evaporation of former isolated bodies of seawater. By the time halite starts to crystallize, only about 1 per cent of the original volume of saline water remains.

Bedded rock salt commonly occurs in layers from millimetres to, in extreme cases, hundreds of metres in thickness. It is generally inter-bedded with other evaporite minerals and with the other sedimentary rocks, especially mudstones, which were laid down during the period of evaporation. Many of the associated mudstones are stained red by iron oxides, and this staining may also be common in the beds of rock salt.

Because rock salt and other evaporite minerals flow easily under pressure, and because the density of salt is generally less than that of overlying sedimentary rocks, bodies of salt may rise from the source layer through the overlying sedimentary rocks. This is especially likely to happen when the initial layer of salt is of uneven thickness. With time the process leads to the formation of a salt intrusion, dome or diapir (Figure 34) which penetrates the overlying and surrounding layers and may even reach the surface. Salt from source layers at great

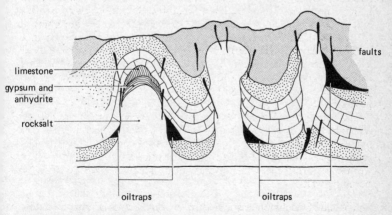

Figure 34.   Salt domes (diapirs) are produced when salt flows as a plastic material into zones of weakness and rises to higher levels in the crust.

depth in folded rocks in Iran has reached the surface, and because of the unusually arid climate it has not dissolved but forms present-day glaciers of salt. Above and around a rising salt diapir a dome may form in the layered sedimentary rocks. Such domes may be traps for oil (♢OIL TRAP).

In addition to salt being mined from bedded deposits, it is also extracted after evaporation from seawater, brines, and some inland SALINE LAKES such as the Great Salt Lake in Utah and the Dead Sea.

In Europe salt deposits are worked at Stassfurt in Germany and in Cheshire, England. The majority of the salt from the Cheshire deposits is obtained from artificially created brines, water having been pumped down to the level of the salt. Extraction of the salt from depth leads to collapse and the development of shallow surface lakes.

The principal use of salt is in chemical industries, the manufacture of GLASS, soap, washing soda, FERTILIZERS, stock feed, weedkiller, paper and ceramics. A small quantity is used for animal and human consumption, and untreated rock salt is used for de-icing roads. Production in 1972, in tonnes, was: USA, 40 843 000; China, 17 820 000; USSR, 12 228 000.

Salt cake. ♢SODIUM CARBONATE AND SULPHATE MINERALS.

Salt diapir. ♢SALT.

Salt dome. ♢SALT.

Sand. A loose, unconsolidated sediment composed of grains between $\frac{1}{16}$ and 2 millimetres in diameter. When consolidated, sand forms the rock type SANDSTONE. The majority of grains in most sands are of the SILICA mineral quartz (SiO_2), but in some sands CALCITE ($CaCO_3$), in the form of shell debris, is predominant. Sands are most abundant in younger sequences mainly of Mesozoic or Cainozoic age. Many sands which were deposited as a result of fluvio-glacial activity are associated with gravel. Some sand is obtained by crushing weakly cemented sandstones and grits.

The uses of sand are many. In terms of tonnage the majority are used for AGGREGATE, but sand is also employed as an ABRASIVE, for making ARTIFICIAL STONE and GLASS, as FOUNDRY SAND, as FILTER SAND, as GRANULES, in HYDRAULIC FRACTURING, as MOULDING SAND, as REFRACTORIES, and in the making of silica bricks.

Sand is worked in association with gravel mainly from deposits of Pleistocene and Recent age. These sands are mainly sharp, that is composed of angular grains. Beach and dune sand is less commonly worked, largely for environmental reasons. Among the so-called 'solid' formations which yield sand in Britain are the Basal Permian sands of County Durham; the Triassic sands of the Midlands; the Jurassic sands of the Lias and the Kellaways Beds; the Wealden, Lower and Upper Greensand sands of Cretaceous age in south-east England; and the Tertiary sands of the London and Hampshire Basins. Many of these deposits yield soft sand, that is sand composed of rounded grains. Where it is poorly cemented the millstone grit of Carboniferous age in the Pennines can be worked for sand.

Shell sands occur abundantly in dune fields along some of the west-facing coasts of Britain, especially in north-west Scotland. Some of them are worked as a source of liming material and as soil conditioners.

Sandfrac sand. ◊HYDRAULIC FRACTURING.

Sandstone. A bedded arenaceous sedimentary rock containing grains between $\frac{1}{16}$ and 2 millimetres in diameter, and generally composed of the mineral quartz (◊SILICA). Sandstones rich in FELDSPAR are called arkoses, those also containing rock fragments and CLAY MINERALS are known as greywackes, and some of those in which MICA flakes are concentrated on bedding planes are flagstones. Some rocks of sandstone texture are formed of detrital CALCITE grains, these rocks being known as calcarenites.

Sandstones are naturally consolidated SANDS. Where the cement is calcite they form calcareous sandstones, where it is quartz they are orthoquartzites, and where the cement is iron oxide they are ferruginous sandstones.

A variety of sandstone which was of great use in construction is flagstone. The rock can be split into thin sheets a centimetre or few centimetres in thickness. It was much used for flagging, roofing and walling.

Ganister is an alkali-deficient sandstone which is of use for refractory purposes. For example, it can be used for FOUNDRY SAND after crushing. Some coarse-grained, rough-to-the-touch sandstones are called GRIT or gritstone.

Sandstones are quarried for a variety of uses but mainly for employment in the construction industry in the form of CRUSHED STONE for AGGREGATE, and as DIMENSION STONE. Sandstones or materials derived from them are also used as ABRASIVES, for making ARTIFICIAL STONE, GLASS, GRANULES and a host of minor industrial uses.

Sapphire. ◊CORUNDUM.

Sassolite. ◊BORATE DEPOSITS.

Saturated zone. ◊ZONE OF SATURATION.

Scandium (Sc). A scarce metal for which there is virtually no commercial demand. Used in research and the production of the radio-isotope, ^{46}Sc. Associated with tungsten, phosphate and uranium ores. World production (1969) was probably about 13·5 kilogrammes.

Scheelite. ⟡TUNGSTEN.

Schist. A metamorphic rock characterized by the development of flat, platy minerals, most commonly micas, which give the fine-scale, layered or foliate texture. Schists are produced commonly from the regional METAMORPHISM of fine-grained sedimentary rocks, less commonly from volcanic beds. Common in the Highlands of Scotland, Scandinavia and the Canadian Shield, also in the Alps and Spain, schists have little but local significance economically, though they may be used as building stone on a small scale.

Schistosity. ⟡CLEAVAGE IN ROCKS.

Seat-earth. ⟡FIRECLAY.

Secondary enrichment. The solution of a metal by surface waters from the upper (leached zone) part of an ore deposit and its redeposition either above (zone of oxidized enrichment) or below (zone of supergene sulphide enrichment) the WATER TABLE.

Sedimentary rocks. Sedimentary rocks are derived from pre-existing rocks as a consequence of weathering, erosion, transport and deposition. They are generally laid down as approximately horizontal layers, or beds, in water, principally in marine, deltaic, estuarine, alluvial or lacustrine environments. Some sedimentary rocks accumulate on the land, for example wind-blown aeolian sands, scree and boulder clay, or till deposited from an ice sheet.

When sedimentary rocks are first deposited they are generally unconsolidated sediments. They become consolidated, or lithified, as a result of natural cementation in the pore spaces, the growth of new minerals in the sediment, and by compaction.

The grains in a sedimentary rock are either clastic (sometimes called detrital) or they are of chemical origin. The detrital grains are those derived, with or without alteration, from pre-existing rocks, whereas those of chemical origin are a consequence of reactions, precipitation following evaporation, or precipitation via the shells and skeletons of animals. Some clastic sedimentary rocks are cemented by chemically precipitated material.

It is difficult to classify all the sedimentary rocks using a single set of criteria. The clastic sedimentary rocks are usually classified on the basis of the size of the grains they contain. The coarsest-grained sedimentary rocks are those containing particles greater than 2 millimetres in diameter. The particles are commonly set in a finer-grained matrix of sand or, less commonly, mud. A typical unconsolidated rock in this category is GRAVEL, a typical consolidated one is CONGLOMERATE. When the average size of the grains is between $\frac{1}{16}$ and 2 millimetres in diameter, the material is of SAND grade. Rocks in this category are

called arenaceous. The unconsolidated sediment is sand and the consolidated rock is called SANDSTONE. The most abundant mineral in
sand and sandstone is quartz ($\lozenge$ SILICA), but a sand-grade rock may
also contain rock fragments, CALCITE and FELDSPAR. The finest
grade of clastic sediment is called MUD or, if a rock, MUDSTONE, and
contains particles of less than $\frac{1}{16}$ millimetre in diameter. These argillaceous rocks, as they have been called, are composed mainly of CLAY
MINERAL particles, rock flour and quartz dust. Well-laminated, fine-
grained sedimentary rocks are called shales.

The remainder of the sedimentary rocks, which are mainly precipitates, either direct chemical precipitates or precipitated via animals or
plants, are generally divided on a chemical basis. In terms of both bulk
and commercial significance the carbonates are of outstanding importance. Those mainly composed of the mineral calcite ($CaCO_3$) are
called LIMESTONE, and those composed primarily of the mineral
dolomite are called DOLOMITES. Rocks rich in ferruginous minerals
such as siderite and chamosite are IRONSTONES. Rocks rich in phosphates are of particular value as PHOSPHATE DEPOSITS. Rocks rich
in chemical silica are called siliceous rocks.

The COALS are rocks predominantly composed of carbon derived
from plants. The relatively rare but economically important rocks
precipitated as a consequence of the evaporation of saline bodies of
water constitute the EVAPORITE DEPOSITS, and contain minerals
which are principally chlorides, nitrates, carbonates and sulphates.

Seepage areas. Relatively slow flow of water through small openings
or porous media to the ground surface. The term is sometimes used to
describe the movement of water in unsaturated soils. More commonly
seepage areas are distinguished from SPRING outlets, the former
indicating a slower movement of ground water to the ground surface.
Water in seepage areas may pond and evaporate or flow depending
upon the size of the seepage, the topography and the climate.

In wells sunk into UNCONFINED AQUIFERS the WATER TABLE may
actually occur slightly above the water level in the well. A free surface
is developed and water can seep into the well between these two levels
through the area known as the seepage face or seepage surface.

Seismic prospecting. Prospecting for natural resources by means of
surface geophysics may involve one of two seismic methods: reflection
or refraction. Seismic reflection methods provide information on geological structures thousands of feet below the surface and are thus a
most valuable tool in the oil industry. On the other hand, in exploration
geophysics, seismic refraction has limited penetrative powers – up to a
few hundred feet in depth – and is of more use in the search for water
resources.

The principle of seismic refraction is based on the fact that shock waves travel through different rock media at different velocities. The denser the material, the faster the waves travel. Seismic or shock waves are produced by setting off an explosive charge in a shallow hole, or by striking the ground with a heavy hammer. A seismograph is an instrument used to detect and time the arrival of the shock wave over a measured distance from the shock point. The velocity of the first part of the shock is then calculated. Comparison of velocities measured at various distances between shock point and detector provides a basis for estimating subsurface geological conditions.

Seismic (shock) wave velocities in unconsolidated, unsaturated material are of the order of 500–1000 metres per second; in unconsolidated aquifers the velocities may vary between 1000 and 2000 metres per second and in AQUICLUDES the velocities can vary between 2000 and 3500 metres per second.

Although essentially a surface geophysical tool, seismic prospecting has been adopted for downhole use. (⟡GEOPHYSICAL EXPLORATION, METHODS.)

Seismograph. ⟡SEISMIC PROSPECTING.

Selenium. Occurs naturally in the form of selenides of copper, gold, lead, silver, mercury, bismuth and platinum. It is used in the manufacture of dry photocopiers (xerography), as a heat-resistent pigment in plastics, in the GLASS industry and in electronics.

Selenium is recovered commercially only as a by-product, mainly during the refining of copper ores. The potential supply from this source is greater than the existing demand. World production (1971) was about 1 million kilogrammes.

Sepiolite. ⟡CLAY MINERALS; MEERSCHAUM.

Serpentine. More correctly serpentenite, it is generally a banded or streaked rock mainly consisting of the serpentine group of minerals. Serpentenites result from the metasomatic alteration (⟡METAMORPHISM) of OLIVINE and PYROXENES in ultrabasic igneous rocks such as PERIDOTITE and dunite. The serpentine minerals are hydrous magnesium silicates. They are soft, and soapy to touch. Serpentenite occurs in some parts of fold mountain belts. Some geologists believe that some of them are altered upper-mantle material.

Some serpentenites are worked for ornamental stone (⟡DIMENSION STONE), as for example on the Lizard Peninsula in south-west England. Some serpentine has been fused with phosphates to make phosphate carriers. They are also of interest because of their association with ASBESTOS and TALC.

Shale. ⟡MUD AND MUDSTONE; SEDIMENTARY ROCKS.

Sharp sand. ⟡AGGREGATE; SAND.

Siderite. Ferrous carbonate containing 48·2 per cent iron, an important iron ore. Distinguished from other carbonates by its brown colour and high SPECIFIC GRAVITY (G = 3·96). Before it can be used in the blast furnace it must be washed or sintered to drive off the contained carbon dioxide.

Sienna. ⟡MINERAL PIGMENTS.

Silica. The different forms of silica (SiO_2) commonly receive special names. Quartz is the crystalline mineral form; it is colourless and transparent when pure, but it is commonly milky white. Its hardness is 7 on Mohs' scale. Quartz is an important constituent of the rock types granite, schist, gneiss and sandstone, and also occurs widely in veins. Gem-quality quartz is known as rock crystal. Cryptocrystalline silica is called chalcedony. Flint and chert nodules, which occur in limestones, are varieties of chalcedonic silica. Concentrically patterned and variegated chalcedony of ornamental value is agate.

As has been suggested already, quartz is an important constituent of many rocks of value. When nearly pure silica is required, as for example in GLASS making, it is generally obtained from pure quartz-rich sands. Silica is widely used as a REFRACTORY material. (⟡ABRASIVES; CEMENT; MINERAL FILLERS.)

Silicate cotton. ⟡MINERAL AND ROCK WOOL.

Siliceous rocks. ⟡SEDIMENTARY ROCKS.

Sill. A variety of sheet intrusion which is parallel to the layering of the sedimentary rocks into which it is intruded (Figure 22) (⟡DYKE). Although most sills are broadly concordant, they jump levels via a dyke in some places. DOLERITE is the commonest igneous rock which forms sills. When a sill is intruded the sedimentary rocks above it have to be lifted by the pressure in the MAGMA, so, not surprisingly, sills are less abundant than the other form of sheet intrusion, a dyke.

Two well-known sills in Britain are the Whin Sill, which underlies much of northern England, and the Central Coalfield Sill in Scotland. Both these sills are of dolerite.

Sillimanite group of minerals. This related group of alumino-silicates yields on calcination (heating) a highly refractory mixture of mullite plus SILICA. Although mullite ($3Al_2O_3 . 2SiO_2$) occurs as a mineral, it is not found in commercial quantities. The three principal source minerals for mullite are andalusite, kyanite and sillimanite, all of which are alumino-silicate (Al_2SiO_5) and occur in metamorphosed, argillaceous sedimentary rocks. Andalusite develops during thermal METAMORPHISM, or low-stress regional metamorphism. It is principally developed in the contact aureoles around intrusive igneous bodies. Andalusite is also present in some PEGMATITES. Kyanite is characteristically developed in moderate to high-grade regional SCHISTS and GNEISSES; it also occurs in pegmatites. Sillimanite is a mineral principally found in schists or gneisses resulting from high-grade regional or thermal metamorphism.

Mullite can also be derived from TOPAZ $(Al(F,OH)_2AlSiO_4)$, a pneumatolytic mineral; dumortierite $(8Al_2O_3.B_2O_3.6SiO_2.H_2O)$, a pneumatolytic or pegmatitic mineral; and pinite, a mica-like alteration product of cordierite, a mineral found in some metamorphosed argillaceous rocks.

Some PLACER DEPOSITS contain concentrations of alumino-silicate minerals, for example, the andalusite sands of the Transvaal.

Mullite is stable at 1110°C and possesses a low coefficient of expansion and a low electrical conductivity. Not surprisingly, it is mainly used as a REFRACTORY material in the metallurgical, ceramic (mullite brick), chemical, electrical, gas and CEMENT industries. It also has a limited use as an ABRASIVE. Kyanite is independently employed in the manufacture of refractory cement, and as a GLASS toughener.

The sillimanite group of minerals is regarded as of strategic importance (⟡STRATEGIC MINERALS), and in recent years entirely synthetic mullite has become increasingly important because many advanced countries have few natural deposits. Production in 1971, in tonnes, of minerals in the group was: USA, 69000; India, 67285; South Africa, 61931.

Silt and siltstone. ⟡MUD AND MUDSTONE.

Silver (Ag). A metal with high malleability and ductility, high electrical and heat conductivity, and resistance to many corrosive agents. Besides its familiar use as silverware and plate, it is used in many electrical appliances and is essential to all forms of conventional photography. Previously used extensively in coinage, most 'silver' coins nowadays are nickel–copper alloys. However, substantial amounts of silver are used for commemorative medallions and other collectors' items; much of the world's silver is hoarded.

Silver (crustal abundance 0·07 ppm) occurs in the native state, as the sulphide (argentite Ag_2S) and in a variety of compounds of arsenic and antimony. Silver may replace part of the crystal lattice of lead and copper minerals. Nearly all commercially produced silver is produced as a co-product or by-product of other, particularly BASE METAL, ores. Canada, the USSR, USA, Mexico and Peru are major producers. Mine production (1974), excluding the USSR, was 240 million troy ounces.

Skutterudite. ⟡COBALT.

Slate. A low-grade, regionally metamorphosed rock. Slates are cut by CLEAVAGE planes which are sufficiently planar and closely spaced to allow the rock to be split into smooth, very thin sheets, generally at an angle to the original layering.

Most slates were originally fine-grained sedimentary or PYRO-CLASTIC ROCKS such as mudstone or ash. The form and origin of cleavage are described under that heading.

Slates are confined to the recent and ancient fold mountain belts (orogenic zones) of the world where significant deformation accompanied folding. Some rocks which can be split into thin sheets parallel to the primary layering (bedding or lamination) are also called slates in the stone trade. They include fissile sandstones and limestones. For example, the so-called Stonefield slate of Middle Jurassic age in the Cotswold Hills of England is a thinly bedded, shelly and quartz-rich limestone which breaks into thin sheets along its bedding planes.

The best slates for roofing and other purposes are those which are strong and cohesive, and which split only along the cleavage. The presence of a second cleavage cutting an earlier slaty cleavage, a not uncommon feature of slates in some orogenic zones, renders the rock unsuitable, because it breaks easily into elongate fragments along two directions. The presence of iron pyrite is also generally considered undesirable because on weathering the iron pyrites breaks down into iron oxides (unsightly), and releases sulphuric acid which may attack the nails holding the slates on to a roof. Calcite in a slate is also undesirable because in polluted areas sulphurous gases may combine with moisture, to form sulphuric acid, which then reacts with the calcite to give gypsum, a material which expands and breaks up the slate.

At present slates are rarely used for roofing except for re-roofing old buildings. They are mainly employed for facing, cladding and ornamental purposes. Thus their colour and texture may be of prime importance. (⟡DIMENSION STONE.)

After quarrying, but before use, slates require splitting and trimming, the latter process resulting in much waste. Spoil tips in the neighbourhood of slate quarries are an unsightly feature of some landscapes. Although slates of commercial grade occur in deposits of considerable bulk, their relative scarcity makes their transport over long distances realistic. Thus they are somewhat intermediate between deposits of low and high monetary value.

Among well-known British slates are the Dalradian dark slates of the Scottish Highlands; the purple and green slates of Cambrian age in North Wales; the dark, commonly blue-black slates of Ordovician and Silurian age in many other parts of upland Wales; the green Ordovician slates, which are cleaved ashes and tuffs, from the English Lake District; and the grey-green Delabole slate of Devonian age in Cornwall.

In the USA most slates are worked in the States of Virginia, Pennsylvania, Vermont and Maine. Elsewhere in Europe, outside Britain, slates are worked in the Ardennes region of France, and in northern Germany. Slate is also worked in Norway, Portugal, Southern Ireland and Italy.

Apart from their constructional uses, some slates belong to the category of EXPANDABLE ROCKS AND MINERALS, and others are used as GRANULES or as fillers (⟡MINERAL FILLERS), as, for example, in the manufacture of some plastics.

Smaltite. ⟡COBALT.

Smithsonite. ⟡ZINC.

Soapstone. ⟡TALC.

Soda ash. ⟡SODIUM CARBONATE AND SULPHATE MINERALS.

Soda-nitre. ⟡NITRATE DEPOSITS.

Sodium carbonate and sulphate minerals. Although there are many sodium-bearing salts, the three evaporite minerals (⟡EVAPORITE DEPOSITS) which are thought of as the sodium minerals are the carbonate, trona ($Na_2CO_3 . NaHCO_3 . 2H_2O$), and the sulphates, mirabilite or natural Glauber's salt ($Na_2SO_4 . 10H_2O$) and thenardite (Na_2SO_4). Other economically important sodium minerals are SALT, nitratine, borax and kernite.

The best-known and most important deposit of trona occurs in rocks of the Green River formation of Eocene age in Wyoming, USA. The trona occurs disseminated in the mudstones of this formation and in beds up to 3 metres in thickness at depths below the surface of up to 500 metres. The conditions of deposition are thought to be those of a former ALKALI LAKE. Where the trona-bearing part of the formation crops out at the surface, the sodium salts are missing on account of solution. The trona mined in Wyoming is converted to soda ash (Na_2CO_3). Sodium carbonate brines are exploited from Searles Lake and Owens Lake in California.

The second most important producer of trona after the USA is Kenya. A crust of trona up to 2 metres thick occurs on Lake Magadi in the East African Rift Valley and natron ($Na_2CO_3 . 10H_2O$), a related carbonate mineral, is dredged from the lake. Hot springs supply soda to the lake faster than it can be removed.

The sodium sulphate minerals are obtained from SALINE LAKE beds in southern Saskatchewan, California and Wyoming. They are also extracted from the lake brines of Searles Lake, and from well brines in Texas. Sodium sulphate deposits also occur on the south-east shore of Great Salt Lake, Utah. Tertiary rocks in the Ebro Basin of Spain contain some sodium sulphate minerals. The sodium sulphate minerals are generally converted to salt cake (Na_2SO_4).

Soda ash is used in the manufacture of caustic soda, sodium bicarbonate and other compounds used in the chemical, GLASS, metallurgical, paper and petroleum industries. Although some soda ash is made from the natural carbonate and sulphate deposits, much of it is manufactured from salt and other raw materials. Salt cake is less widely used than soda ash but it is employed in the manufacture of brown paper, and in the glass, textile, dye and photographic industries. Most salt cake is obtained as a by-product in the manufacture of hydrochloric and sulphuric acid.

Production of sodium carbonate in 1971, in tonnes, was: USA, 2611000; Kenya, 161260; for sodium sulphate it was: USA, 614000; Canda, 437190.

Soft sand. ⟢AGGREGATES; SAND.

Soil. In general terms, the upper layer of weathered or transported loose rock and mineral fragments in which plants grow and to which they contribute humus. Soils constitute valuable resources because they are the basis for agriculture. They may be 'improved' by IRRIGATION and chemical or other fertilizing agencies (⟢FERTILIZER SOURCES) and they may be 'destroyed' by excessive cropping, flooding or contamination by natural salts or industrial pollutants.

H. O. Buckman and N. C. Brady, *The Nature and Properties of Soils*, Macmillan, 7th edn, 1969.

Soil water zone. Water in the soil water zone or subzone exists at less than saturation except temporarily, when excessive water reaches the ground surface as from rainfall or irrigation. The zone extends from the ground surface down through the major root zone, and its thickness varies with soil type and vegetation.

Soil water comprises three types dependent upon concentration in the soil zone. HYGROSCOPIC WATER, adsorbed from the air, forms thin films of moisture on soil particle surfaces. This water is unavailable to plants. Capillary water exists as continuous films around soil particles. This water is held by surface tension yet moves by capillary action and is, therefore, available to plants. GRAVITATIONAL WATER is excess soil water which drains through the soil under the action of gravity.

Soil moisture studies represent efforts to define the types of soil water. For example capillary water, based on moisture content, occurs between the hygroscopic coefficient (⟢HYGROSCOPIC WATER) and FIELD CAPACITY.

Sonde. Miniaturized geophysical equipment capable of being lowered down a well. In ELECTRICAL LOGGING of wells two important methods are used, namely normal sonde and lateral sonde. In the former, two electrodes are lowered down the well with two more fixed at the surface. In the lateral sonde technique three electrodes are lowered downhole with one fixed at the surface. The sonde spacing is the distance between the measuring and current electrodes. (◊GEO-PHYSICAL EXPLORATION, METHODS.)

Sonic logs. A sonic or acoustic log records the speed of sound in the rocks near a well. Because the travel time of the sound is strongly dependent on the porosity, the sonic log is widely used to estimate POROSITY of RESERVOIR ROCKS. There has been successful application of this technique to moderately porous sandstones, limestones and dolomites. There have been more recent developments in the application of the technique to locating fractures and fissures in dense, consolidated rocks.

Source rock, petroleum. The original formation to contain the organic material from which the HYDROCARBONS of oil and gas are derived (◊PETROLEUM). Most commonly these are shallow-water, dark marine limestones and shales. The organic material is the debris of once-living organisms alive when the sediment was forming. The transformation of this into hydrocarbons requires heat, pressure and time. Source rocks may be of Palaeozoic, Mesozoic or Cainozoic age. Many have a petroliferous smell when freshly broken. Precambrian strata rarely contain source rocks for oil or gas.

S.P. curve. ◊ELECTRICAL LOGGING.

Specific capacity. The specific capacity of a well is its yield per unit of DRAWDOWN, usually expressed as gallons per hour, or per minute, per foot of drawdown. Dividing the yield from the PRODUCTION WELL by the drawdown in that well, each measured at the same time, gives the value of specific capacity. Specific capacity measurements can be used in approximating the TRANSMISSIVITY of the aquifer.

Specific gravity. The ratio of an object's or substance's weight to that of an equal volume of water. The property is useful in the field identification of certain minerals, commonly those which have a high specific gravity, e.g. barites 4·5; galena 7·4–7·6; pitchblende or uraninite 6·4–9·7.

Specific retention. The ratio expressed as a percentage of the volume of water that a rock or soil will retain after saturation against the force of gravity to its own volume. Specific retention plus specific yield (◊EFFECTIVE POROSITY) equals the POROSITY of the rock or soil sample. Specific retention, therefore, relates to the volume of water retained by molecular attraction in the sample and is linked to FIELD CAPACITY, PELLICULAR WATER and RETAINED WATER.

Specific yield. ◊EFFECTIVE POROSITY.

Sphalerite. ⬦ZINC.

Spinel. MAGNETITE (Fe_3O_4) and CHROMITE ($FeCr_2O_4$) are minerals of economic importance which belong to the spinel group. Crystals of minerals belonging to the group are cubic.

Spodumene. ⬦PYROXENE.

Springs. Concentrated discharges of ground water appearing at the ground surface as a current of flowing water (Figure 3). In addition to appearing at ground surface, springs may break out directly into a body of inland water, such as a river or lake, or into the sea. Springs appear where the WATER TABLE intersects the surface or where a fissure or fracture allows the emergence of water.

Springs may be classified in a number of ways. Classifications have been based on magnitude of discharge, type of aquifer, chemical characteristics, water temperature, relation to topography and geological structure. A simple classification is into springs which result from gravitational or non-gravitational forces.

Examples of gravity springs occur where surface topographic depressions in permeable rocks extend down to the WATER TABLE, as in many valleys. Others occur where water flows from permeable rock strata at the exposed junction with underlying, relatively impermeable beds, or at the foot of an escarpment of a permeable rock formation. A similar type of gravity spring may occur where the permeable rock formation is saturated, throughout its entire thickness, up to the exposed contact with the overlying impermeable beds.

Non-gravitational springs include volcanic and fissure springs, the latter associated with deep crustal fractures. These types of spring are usually thermal in origin in that their water temperature exceeds that of 'normal' ground water. The terms warm and hot springs are frequently used. Waters of thermal springs are commonly highly mineralized and may consist solely of JUVENILE WATER or a mixed meteoric-juvenile water.

Steatite. ⬦TALC.

Stibnite. ⬦ANTIMONY.

Stockpile. Ore mined and accumulated for future use. Ore may be stockpiled for only a short period, perhaps while awaiting shipment or processing, or may be deliberately stockpiled for a long period of time. Most governments stockpile STRATEGIC MINERALS, imported if necessary, against times of crisis or emergency. The existence of large stockpiles can influence metal prices if it is threatened to release ore on the market.

Stone. Apart from its everyday use to describe a small piece of rock, the word stone is employed in the quarrying and mining industries to cover any hard rock. Some gems are also called stones.

Storage coefficient. The volume of water an aquifer releases from or takes into storage per unit surface area of the aquifer per unit change in head. From this definition it can be seen that the storage coefficient is dimensionless.

The storage coefficient of UNCONFINED AQUIFERS is virtually equal to the specific yield ($\diamond$ EFFECTIVE POROSITY), as most of the water is released from storage by gravity drainage and only a small part comes from compression of the aquifer and expansion of the water.

The storage coefficient of CONFINED (ARTESIAN) AQUIFERS is related to elastic compression of the aquifer when water is withdrawn from it through a PRODUCTION WELL. Withdrawal of water from a confined aquifer results in a decline in artesian pressure and a subsequent increase in the load supported by the aquifer. Thus the aquifer undergoes elastic compression or compaction with the removal of water from storage. In addition to storage discharge, there is additional supply of water by the expansion of the water itself through the reduction in pressure. Confined aquifers thus exhibit volume elasticity, and pressure can be restored in these aquifers through recharge with recovery of water levels, expansion of the aquifer and compression of the water within the aquifer.

The storage coefficient of most confined aquifers ranges from about 10^{-5} to 10^{-3} and is about 10^{-6} per foot of thickness. In contrast the storage coefficient (specific yield) of most unconfined aquifers ranges from about 0.1 to 0.3.

Strategic mineral. A mineral which is considered to be essential to industrial activity in a country especially during wartime. Most strategic minerals are relatively rare, and they are not abundant in all countries. Thus it is common for many nations to stockpile strategic minerals. Among the minerals or rocks regarded as of strategic value by DeMille in 1947 are those in the SILLIMANITE GROUP, and those which are sources of chromium, manganese, aluminium and tin.

Stratiform ore deposit. An ORE (1, 2) deposit (especially of copper, lead and zinc) confined to individual sedimentary horizons. The mineralization is generally assumed to have taken place at about the same time as the formation of the sediments ($\diamond$ SYNGENETIC DEPOSIT). Notable examples of stratiform deposits occur in the thin 'Kupferschiefer' of East Germany and Poland, and the very rich Proterozoic sediments of the Zambian copper belt.

Strike. The strike of a surface, such as a bedding plane, in a rock mass is the direction of any imaginary horizontal line on that surface. The direction of strike is always at right angles to the direction of DIP (Figure 19).

Strontianite. ◇STRONTIUM.

Strontium (Sr). Used mainly as strontium compounds in pyrotechnics (giving an intense red or crimson colour) and ammunition. It is also used in the manufacture of high-purity electrolytic zinc, as an additive to grease and in ceramic glazes. Strontium hydroxide is used in the production of sugar from beet sugar.

Celestite ($SrSO_4$) and strontianite ($SrCO_3$) are the main ore minerals and occur in veins, segregations and pockets often associated with EVAPORITE DEPOSITS.

Reliable production statistics are unobtainable but the major producers in 1971 were Mexico, Canada and the United Kingdom (9749 tonnes). The United Kingdom production is from celestite deposits which occur in the Keuper marl near Bristol.

Structural clay products. CLAYS which are used in the manufacture of bricks (◇BRICK CLAYS), pipes, tiles (◇PIPE AND TILE CLAYS) and other related, relatively low-grade products are the principal raw materials employed for this group. Most of the clays used are plastic; yield material of relatively high strength when fired; and vitrify, that is form glass, over a fairly broad range of temperatures. They contain a higher proportion, and a wider range of impurities, such as iron oxides, organic material, quartz and calcite, than clays used for higher-grade products.

In most parts of the world there are clays, mudstones and shales which are used for making bricks, tiles and pipes. Their large bulk and low monetary or unit values mean that they are rarely exported.

Subsurface water. In the normal sequence of the HYDROLOGICAL CYCLE, that part of precipitation which is not evaporated or which does not flow away as surface RUNOFF, penetrates into the ground and thereby enters the province of subsurface water. Subsurface water comprises the various waters present in the AERATION ZONE – vadose or suspended waters – together with the water in the ZONE OF SATURATION – ground (phreatic) water (Figure 7).

Sulphate. Sulphate in ground water is derived principally from the EVAPORITE DEPOSITS gypsum and anhydrite. Another source is precipitation where the sulphate content may be as high as 10 ppm. It may also be derived from the oxidation of metallic sulphides, e.g. pyrite. Magnesium sulphate (Epsom salt) and sodium sulphate (Glauber's salt), if present in sufficient quantities, will impart a bitter taste to the water, and the water may act as a laxative for people not accustomed to drinking it.

High sulphate levels in ground waters of COASTAL AQUIFERS are associated with SALINE INTRUSION. Away from coastal influences local high sulphate content in aquifers may be due to pollution, possibly with FERTILIZERS as the source. However, the concentration of the sulphate ion in natural waters may be affected by the life processes of sulphate-reducing bacteria. These bacteria derive energy from the oxidation of organic compounds and in the process obtain oxygen from the sulphate ions in ground water. The process of sulphate reduction by these bacteria is to break down the sulphates to sulphides. This reduction of sulphate ions produces the gases hydrogen sulphide and carbon dioxide.

Sulphate-reducing bacteria have been recognized as indigenous in extremely saline, low-sulphate waters associated with oilfield brines. These bacteria may be indigenous in ground waters and the smell of hydrogen sulphide ('bad eggs') gas is detectable at many well heads, particularly wells drilled into CONFINED AQUIFERS. Corrosion of well CASING in such situations is common, and is often ascribed to these bacteria.

Sulphur (S). The bright yellow, non-metallic element sulphur is used mainly in chemical processes. The three principal sources of sulphur and sulphur compounds are accumulations of native elemental sulphur (brimstone), several of the heavy metal sulphide minerals such as iron pyrite, copper pyrite and pyrrhotite, and sulphates such as anhydrite. Sulphur is also obtained from hydrogen sulphite, a by-product of oil and gas processing.

Native sulphur occurs in association with some active volcanic centres and fumaroles. Volcanic sulphur is worked in Japan, and considerable quantities are reported to be available in the Andes Mountains. Sulphur deposits in some limestones may be the result of hot sulphurous waters having reacted with the limestones. Sulphur replaces gypsum or anhydrite in some EVAPORITE DEPOSITS. Perhaps the best-known occurrence of native sulphur is in the CAP ROCKS above a SALT diapir, or intrusion, in an oilfield. Where sulphur of this type occurs in the Gulf States of the USA it is mined by the Frasch process. Many theories have been advanced to explain the presence of sulphur in the cap rocks above salt intrusions. One favoured idea is that it develops as a consequence of the activity of sulphur-reducing bacteria on anhydrite.

The majority of European sulphur is obtained from iron PYRITE (FeS_2) and to a lesser extent from copper pyrite and pyrrhotite. Iron pyrite occurs in commercial quantities in some hydrothermal veins. The Rio Tinto mines in south-west Spain yield about 25 per cent of world production. They and similar deposits in Cyprus have been worked since Phoenician times. The extraction of copper, gold and arsenic as by-products make the working of pyrite for sulphur a profitable enterprise.

Sulphur is principally employed in the manufacture of sulphuric acid. It is also used in ferrous and non-ferrous metallurgical processes, in the manufacture of paint, rayon, cellulose, explosives, insecticides, fungicides, rubber, paper and acid-proof CEMENT. Production of elemental sulphur in 1972, in tonnes, was: USA, 9300000; Canada, 6900000; Poland, 3100000.

Supergene enrichment. ◊SECONDARY ENRICHMENT.

Syenite. A coarse-grained, plutonic igneous rock of intermediate composition. Syenites commonly contain alkali FELDSPAR, hornblende and less than 10 per cent of quartz. Some syenites which are silica-deficient contain FELDSPATHOID MINERALS such as nepheline. A nepheline syenite in Ontario, Canada, is quarried and used, after treatment, as a substitute for feldspar in the ceramics industry.

Elsewhere some syenites are of value for CRUSHED STONE. An especially coarse-grained syenite known as larvikite, from Larvik in Norway, is widely used as a facing stone because of the remarkable iridescence shown by the feldspar crystals that it contains. (◊DIMENSION STONE.)

Sylvinite. ◊POTASH DEPOSITS.

Sylvite. ◊POTASH DEPOSITS.

Syncline. ◊FOLD.

Syngenetic deposit. A term applied to a mineral or ore deposit formed at the same time as the enclosing rocks. The opposite of EPIGENETIC DEPOSIT.

T

Taconite. ⟡BANDED IRON FORMATIONS.

Tailings. GANGUE and other material that cannot be economically recovered following the washing, concentration or treatment of ground ORE: the fine-grained residual waste which is rejected after the mining and processing of ore. At a later date, after a rise in price or an improvement in recovery techniques, tailings may become of economic value.

Talc. A white, greasy-feeling mineral $(H_2Mg_3(SiO_3)_4)$ which is point 1 on Mohs' scale of hardness. It is a major constituent of the rock type soapstone or steatite. Talc is commonly associated with the hydrothermal alteration of ultrabasic igneous rocks, in which it occurs as veins associated with shear zones. Talc also results from the low-grade thermal metamorphism of some sandy dolomites. This association generally yields the talc of best commercial quality.

Talc is used as an extender or MINERAL FILLER in the manufacture of many products. It is also employed as a soft ABRASIVE, in medicine and cosmetics, and as a lubricant, for example french chalk. Production in 1971, in tonnes, was: Japan, 1895000; USA 849363; USSR, 400000. (⟡GRANULES.)

Tantalum (Ta). A minor metallic element used principally in the electronics industry for the manufacture of capacitors and rectifiers, and in the production of corrosion-resistant alloys for the chemical industry. A minor but important use is in surgery (e.g. pins for fractured bones), where tantalum is characterized by its inertness to body fluids and its tolerance by the human body.

Tantalum is obtained from the columbite–tantalite mineral series which is an isomorphous series with the end members $(FeMn)Ta_2O_6$ and $(FeMn)Cb_2O_6$. The minerals are associated with granite PEGMATITES but are frequently found in PLACER DEPOSITS.

The USSR, Brazil, Australia, Mozambique and Zaire are producers. World production (metal content) in 1968 was an estimated 2·4 million lb.

Tar. ⟡ASPHALT.

Tar sands. Sedimentary formations, largely of sand, sandstone or siltstone, in which the bulk of the cementing material between the mineral grains is thick tar. The tar is made of heavy, large-molecule, asphaltic HYDROCARBONS. In warm weather tar sands may 'weep' droplets of oil but they do not flow.

Largest of all tar sand formations are the Athabaska tar sands covering 48 000 square kilometres of northern Alberta, Canada, and as much as 90 metres thick. They are said to contain the equivalent of 600×10^9 barrels of oil but production on an economic basis is not yet developed.

Tellurium (Te). One of the rarer elements (crustal abundance 0·01 ppm). It occurs naturally as tellurides and is found in copper, lead, gold and other ores (cf. selenides of SELENIUM). Used mainly as an additive to steels, it is recovered commercially only as a by-product during the refining of copper and lead ores. World production (1974) was of the order of 400 000 lb.

Temperature logs. One of the most conservative properties of ground water is temperature. Under normal conditions the temperature of ground water will vary, within a few degrees, around 52°F at usual shallow ground water abstraction depths. Below shallow depths the temperature increases approximately 1°C for each 30 metres of depth in accordance with the heat gradient of the earth's crust.

Downhole examination of water wells measuring ground water temperatures can be made with a recording resistance thermometer. Derived data may be invaluable in analysing subsurface conditions. Departure from the normal thermal gradient, mentioned above, may provide information on water circulation or geological conditions in the well. Marked temperature variations may indicate water from different aquifers penetrated by the well. Temperature variations may also be produced by entry into the well of waters from major fissures.

Tenor. The amount of valuable metal in an ore. Usually given as the percentage of metal or metal oxide in the ore. The tenor of precious metals is given as troy ounces per avoirdupois ton (1 ton = 29 167 troy ounces) or in pennyweight (dwt) per ton (1 troy ounce = 20 dwt).

Test drilling. Drilling of small-diameter holes to determine the geology of areas potentially productive of water, minerals or valuable rock materials. In the search for water it is to ascertain geological and ground water conditions at any site prior to the drilling of the main PRODUCTION WELL. Many times, if a test hole proves fruitful, it is redrilled, or reamed out, to a larger diameter to form a production well. Test holes may also serve as OBSERVATION WELLS for measuring water levels.

In test hole drilling the primary things to be obtained include the identification and location of the site of each test hole, the log of the rock strata penetrated (well log) with representative samples of these strata, and samples of water from possible aquifers for quality determination.

Almost any well drilling method can be employed in test drilling but in unconsolidated formations cable-tool and hydraulic rotary are most common (⇨DRILLING METHODS). A drilling time log is a useful supplement to test drilling. It consists of an accurate record of the time, in minutes and seconds, required to drill each foot of the well. This can provide valuable information on the type of strata penetrated, because variations in rock texture largely control the drilling rate.

Thallium (Tl). An extremely toxic element used principally in the electronics industry. It is recovered entirely as a by-product during the processing of BASE METAL (especially zinc) ores.

Thenardite. ⇨SODIUM CARBONATE AND SULPHATE MINERALS.

Thixotropy. Thixotropic materials are colloids which are capable of changing from a gel to a sol when agitated, but reverting to a gel when left to stand. Some bentonitic clays (⇨BENTONITES) are thixotropic when mixed with water, a property which is utilized when they are employed in DRILLING MUDS.

Thrusts. ⇨FAULT.

Tin (Sn). Utilized by man, as an alloy (metal) in bronze, since prehistoric times. Nowadays the main uses are for coating steel to make tin plate (especially for food cans), the production of solders, and in various alloys. It is an expensive metal; prices in 1973 attained £2700 per tonne, so considerable efforts are made to recycle tin. Although tin (crustal abundance 2 ppm) occurs in complex sulphides, the oxide CASSITERITE (SnO_2) is the only important ore of tin. Primary deposits, as LODES and replacements, are genetically associated with granitic rocks. The main commercial sources, however, are PLACER and ELUVIAL DEPOSITS.

World mine production in 1971 was an estimated 234000 tonnes, of which half was derived from the remarkable METALLOGENIC PROVINCE extending through Thailand (9 per cent of world total), West Malaysia (32 per cent) and Indonesia (9 per cent). Significant production also came from Bolivia (13 per cent, mainly vein deposits), the USSR (11 per cent) and China (9 per cent). The USA, rich in so many minerals, had virtually no commercial production. Some 3500 tonnes of tin were mined in Cornwall in 1973 – small by world standards but an important contribution to Britain's requirements.

Titania. ⇨TITANIUM.

Titanium (Ti). The ninth commonest element in the earth's crust. Over 90 per cent of the titanium produced is used in the form of the oxide, titania, as a white pigment or MINERAL FILLER in the paint, paper and plastic industries. Titanium metal and its alloys have a favourable weight-to-strength ratio combined with a high resistance to corrosion; they are being increasingly used in the aerospace technologies.

Only two minerals, ilmenite ($FeTiO_3$) and RUTILE (TiO_2), are of commercial importance. Both minerals are common constituents of PLACER DEPOSITS, and the bulk of rutile comes from such deposits. Ilmenite also occurs in segregations, usually associated with other iron ores, in gabbroic rocks (especially a feldspar-rich variety known as anorthosite).

World production of ilmenite (excluding the USSR) in 1971 was 4·9 million tonnes. The leading producers are Australia, the USA, Canada, the USSR (but figures not available) and Norway. Production of rutile, for technological reasons the preferred source of titanium metal, was only 384 000 tonnes, and the bulk of this (367 000 tonnes) came from Australian placer deposits.

Ton. The British (UK) ton is an old unit of weight equivalent to 20 hundredweights or 2240 pounds, avoirdupois measure. In the USA it is called the long ton or gross ton; the American equivalent is the short ton or net ton, 0·893 × the UK ton. The metric equivalent the tonne (tonneau) or metric ton, is equal to 1000 kilogrammes.

Topaz. The orthorhombic mineral topaz ($|Al(F,OH)|_2SiO_4$) is commonly included with the SILLIMANITE GROUP OF MINERALS because, after processing, it is used as a REFRACTORY material. Topaz occurs in pneumatolytic PEGMATITES and veins associated with granite intrusions. It has been worked in South Carolina and India. Some topaz is of gem quality.

Total stock. The amount of a RESOURCE as determined by its geophysical and geological distribution within the earth. Only a portion of a total resource stock is ever likely to become a RESERVE.

Tourmaline. This mineral, a complex sodium, lithium, iron, aluminium hydroxyl borosilicate, is found in some PEGMATITES and veins associated with granites. Some tourmaline is of gem quality and some varieties have been used for piezoelectric components, and for obtaining polarized light.

Tracers. Commonly employed to follow the movement of ground water through aquifers, in order to establish both the direction and velocity of ground water flow. Methods using dyes and salts have been used for many years and are best applied to aquifers possessing a ramifying network of fissures in which fissure flow is dominant. With intergranular flow in a porous medium, tracers are liable to be dispersed by the microvelocity variations inherent in LAMINAR FLOW through pores.

Ideally tracers should be quantitatively determinable in very low concentrations. They should be absent or nearly absent from the natural water and must not react with the natural water to form a precipitate. Tracers must not be absorbed by porous media and must be readily available and inexpensive.

Tracers may be classified as to method of detection, namely, colourmetric, chemical determination, radioactive, electrical conductivity, etc. Certain radio-isotopic tracers, e.g. tritium, can be used as field tracers without danger of contamination, but others must be carefully controlled because of dangerous radiation levels.

Trachyte. A fine-grained, generally pale-coloured, volcanic igneous rock of intermediate composition. It is said to be a rock which is rough to the touch. Trachytes contain alkali FELDSPAR and soda-rich AMPHIBOLES or PYROXENES. Many possess lath-like larger crystals (phenocrysts) of feldspar arranged parallel to each other and the presumed flow direction of the lava. In composition trachytes are equivalent to SYENITE. Some trachytes are worked for CRUSHED STONE.

Transmissivity. The transmissivity of an aquifer is the product of the HYDRAULIC CONDUCTIVITY (permeability) and the aquifer thickness. Its usefulness in aquifer studies is paramount in that it measures the movement of ground water through the aquifer, particularly in the vicinity of PRODUCTION WELLS.

The transmissivity of an aquifer can be defined as the rate at which ground water flows through a vertical strip of the aquifer one foot wide, extending through the full saturated thickness of the aquifer, under a hydraulic gradient of 1 or 100 per cent. In practice transmissivity is normally measured in gallons per day per foot or square metres per day.

Values for transmissivity range from less than 1000 to over 1 000 000 gallons per day per foot (12 ◊ 12 000 m²/d). An aquifer of transmissivity less than 1000 can supply limited amounts of water to small domestic wells. Where the transmissivity is of the order of 10 000 or more, well yield may be adequate for municipal, agricultural (irrigation) or industrial purposes.

Transmissivity and STORAGE COEFFICIENT (of aquifers) are of fundamental importance in that they define the hydraulic characteristics of the water-bearing formation. These two parameters can be gauged by PUMPING TESTS on wells. The transmissivity indicates how much water moves through the aquifer and the storage coefficient reveals how much ground water can be removed by pumping or draining. If these two parameters are known for a particular aquifer, significant predictions in water resource management can be made. (◊SPECIFIC CAPACITY.)

Transpiration. Refers to the EVAPORATION of water absorbed by plants. It is a process by which plants lose water to the atmosphere. In many regions the overall evaporation cannot be measured separately from transpiration and the two effects are considered together as evapotranspiration.

Travertine. ◊LIMESTONE.

Tremolite. ◊AMPHIBOLE; ASBESTOS.

Tripoli. A white-to-red, low-density, friable, porous silica-rich material which results from the decomposition of siliceous limestones. It is not to be confused with tripolite, an obsolete term for DIATO-MITE from Tripoli in Libya. Tripoli is worked in several eastern and mid-western states in America, and in Spain, Portugal and Australia. In Britain certain rottenstones (decalcified limestones), which are now rich in silica, have been exploited for tripoli-like material.

The principal use of tripoli is as an ABRASIVE, especially for buffing and polishing compounds. Some tripoli has also been used as a MINERAL FILLER in paint and other products.

Tripolite. ◊DIATOMITE; TRIPOLI.

Trona. ◊SODIUM CARBONATE AND SULPHATE MINERALS.

Tufa. ◊LIMESTONE.

Tuff. ◊PYROCLASTIC ROCKS.

Tungsten (W). A heavy, hard, heat-resistant metal used principally as tungsten carbide (for cutting tools), in special steels and non-ferrous alloys. A familiar use is as a filament in electric light bulbs. Wolframite ($FeWO_4$) and scheelite ($CaWO_4$) are the most important tungsten minerals. They occur in quartz veins and contact metamorphic deposits (especially scheelite) associated with granitic rocks, and in PLACER DEPOSITS. Mainland China is a major producer of tungsten. Estimated world production (1974) was in the order of 44000 tonnes.

Turbulent flow. Turbulent flow conditions occur when small, random velocity fluctuations occur across the line of fluid flow. These cross fluctuations are caused by eddies generated as the fluid moves past obstacles or along rough boundaries. A small particle suspended in turbulent flow does not follow a smooth path, as in LAMINAR FLOW, but wanders up and down and from side to side as it moves along.

Turbulent flow conditions exist where the velocities exceed the restriction by viscous forces. This is the usual condition across almost the whole section of most natural surface streams and is also present in the immediate vicinity of large-diameter water wells pumping at high discharge rates. Turbulent flow conditions are also frequently present in KARST aquifers where large solution cavities contain ground water moving at relatively high velocities.

U

Ulexite. ◊ BORATE DEPOSITS.

Ultrabasic (ultramafic) rocks. ◊ IGNEOUS ROCKS.

Umber. ◊ MINERAL PIGMENTS.

Unconfined aquifer. An AQUIFER which contains water that is in direct contact vertically with the atmosphere through open spaces in permeable material. The water contained therein is termed unconfined water.

Thus ground water in unconfined aquifers occurs under WATER TABLE conditions. This means that the upper limit of ground water in such an aquifer is defined by the water table itself (Figure 6). The hydraulic pressure at any level within an unconfined aquifer is equal to the depth from the water table to the point in question and may be expressed as hydraulic head in feet of water. For example, ground water at a depth of 30 metres below the water table is under a hydraulic head of 30 metres.

When a well is drilled into an unconfined aquifer the rest water level in the well stands at the same elevation as the water table. Thus the water table can be determined by an examination of water levels measured in a series of wells sunk into the unconfined aquifer. Wells drilled into an unconfined aquifer are termed water table wells, the rest levels of which will fluctuate depending upon recharge to and discharge from the aquifer.

A PERCHED AQUIFER is a special case of an unconfined aquifer occurring when a ground water body is separated from the main ZONE OF SATURATION by a rock unit of relatively low permeability and limited areal extent within the AERATION ZONE (Figure 7). (◊ CONFINED (ARTESIAN) AQUIFER.)

Underclay. ◊ FIRECLAY.

Underflow. Ground water flow at depth through an aquifer. Underflow may occur under either UNCONFINED or CONFINED AQUIFER conditions. This deep-seated flow through aquifers is important in that

187

it must be considered in any hydrological balance undertaken over a ground water catchment.

Unit value. ◊MONETARY OR UNIT VALUE.

Unsaturated zone. That part of an UNCONFINED AQUIFER where VADOSE WATER only partially fills the INTERSTICES. Thus the unsaturated zone exists everywhere above the WATER TABLE (Figure 7). It is synonymous with the AERATION ZONE, comprising three sub-zones, and where two-phase flow occurs through the interstices only partially filled with water and air. (◊ZONE OF SATURATION.)

Uraninite. This important source of URANIUM is the mineral uranium dioxide (or pitchblende), occurring in a wide variety of rocks both crystalline and sedimentary. It has been found widespread but in very low concentrations in rocks in North America, Africa, Australia and central Europe.

Uranium (U). The principal radioactive element used in the production of nuclear (or radioactive) energy. Uranium has several similar kinds of atoms, but only the natural kind known as ^{235}U is 'radioactive' or unstable. Uranium-235 comprises only about 0·7 per cent of all natural uranium, the rest being largely ^{238}U which can only be used as a nuclear fuel after treatment.

Uranium-235 breaks down into lighter elements, neutrons and heat in what is called a fission reaction. The 238 form can be converted into the active element plutonium-239 by the process known as breeding and the form of the element thorium, known as thorium-232, can be converted into fissionable ^{233}U by a breeder reaction.

It is calculated that 1 kilogramme of uranium produces heat energy equivalent to about 11 800 barrels of crude oil.

Uranium occurs in a variety of minerals, especially in veins associated with granite-like and other crystalline rocks and in certain sedimentary rocks (◊URANINITE). Large deposits are found in Canada, western USA, Australia and South Africa (Figures 35 and 36). (◊BACTERIAL EXTRACTION OF METALS.)

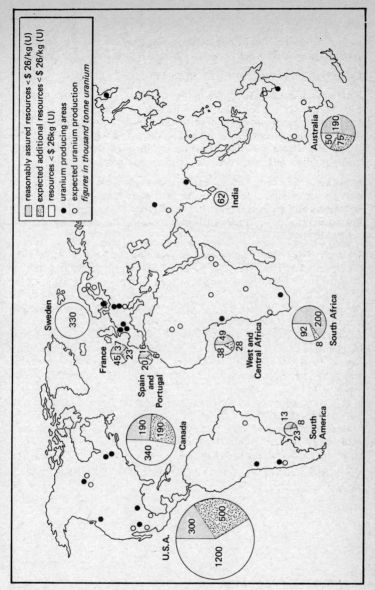

Figure 35. World distribution of uranium producing areas and localities which may be expected to produce uranium in the future. (After *Nuclear Engineering International*.)

189

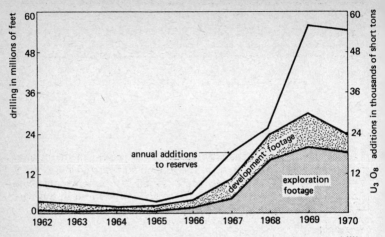

Figure 36. Reserve addition and exploration and development drilling for uranium minerals.

V

Vadose water. Subsurface water which is suspended in the unsaturated AERATION ZONE between the ground surface and the WATER TABLE in UNCONFINED AQUIFERS (Figure 7); synonymous with suspended water, it includes the various waters contained in the SOIL WATER ZONE, INTERMEDIATE ZONE and CAPILLARY FRINGE.

Vanadinite. ◊VANADIUM.

Vanadium (V). Used principally as an alloying element with iron and steel to increase strength, ductility and resilience. Vanadium compounds are used as catalysts in the chemical industry.

Although not a scarce element (crustal abundance 135 ppm), vanadium minerals rarely form commercial deposits themselves. The principal minerals are carnotite (hydrous potassium–uranium vanadate), roscoelite (vanadium mica), vanadinite and descloizite (complex lead–vanadium minerals). They occur in diverse environments: examples include the 'red-bed' copper–vanadium–uranium deposits of the western USA, the vanadium minerals found in the oxidized zone (◊SECONDARY ENRICHMENT) of base metal ores of south-west Africa, and the vanadiferous titaniferous magnetite of the Bushveld complex of South Africa. Venezuelan crude oil and the Athabasca TAR SANDS of Canada are important potential sources of vanadium.

Free-world production (1973) in terms of metal content was an estimated 17500 tonnes, with South Africa and the USA the leading producers. Russia is also a major producer.

Vein. A zone of mineralized rock usually sharply delineated from the surrounding 'country rock'. Veins are usually narrow, often irregular, and steep or vertical in aspect. They often contain high-grade but limited amounts of ore; in the past they were the main source of many metals. LODE is commonly used as a synonym.

Vermiculite. ◊CLAY MINERALS; EXPANDABLE ROCKS AND MINERALS.

Vitrification. Material which is capable of vitrifying forms a glassy, non-crystalline solid when heated into its vitrification range. Many

CLAYS vitrify when heated to high temperatures, a property made use of in ceramics and brick making. (◊ BRICK CLAYS; CERAMIC CLAYS; STRUCTURAL CLAY PRODUCTS.)

Voids. The open spaces between solid material in a porous medium. Voids, PORES and INTERSTICES are terms which are closely interrelated and in part synonymous. Vugular pore space relates to void spaces in a porous medium produced by solution cavities of a small size. (◊ POROSITY.)

Volcanoes and volcanic rocks. ◊ IGNEOUS ROCKS.

Vugular pore space. ◊ VOIDS.

Wad. A dark brown or black, earthy mixture of manganese and other oxides. Usually contains iron oxides and occasionally barium and cobalt (variety: asbolite).

Water analysis. Once a ground water sample has been collected, it is normally subjected to a chemical analysis in the laboratory. Ground water contains dissolved minerals taken into solution as it percolates through the soil or rock media. Most ground water contains no suspended matter and practically no bacteria. It is usually clear and colourless, and it may contain dissolved gases. These characteristics contrast sharply with those of most surface waters, which are usually turbid and contain considerable quantities of bacteria.

The intended use of the ground water controls in large part the number of constituents determined in a water analysis. More than fifty properties are subject to determination but this would be a very complete analysis, in excess of the normal analysis which would provide a picture of the ground water for the usual domestic, municipal, industrial or agricultural use.

In a water analysis concentrations of different ions are expressed by weight or by chemical equivalence. Total dissolved solids can be measured in terms of electrical conductivity. Many laboratories report analyses in units of milligrammes per litre, which are usually regarded as being the same as parts per million (ppm). Analyses are also expressed in equivalents per million (epm) which are calculated by dividing the concentration value of a substance, in ppm, by its combining weight. The combining weight equals the atomic or molecular weight of the ion divided by its ionic charge.

Analyses of ground waters may be in the form of a partial analysis or full mineral analysis, both of which contain an analysis of the physical properties of the water, together with a bacterial analysis. The ions commonly determined in a water analysis are calcium, magnesium, sodium, potassium, carbonate, bicarbonate, sulphate, chloride and nitrate. The total dissolved solids, hardness, alkalinity and pH are also often included in the water analysis.

Water spreading. The releasing of water over the ground surface in order to increase the quantity infiltrating into the ground, through the AERATION ZONE to the WATER TABLE. It is, therefore, a method of ARTIFICIAL RECHARGE of ground water. The rate at which water will enter the soil largely depends upon the area of recharge and the time water is in contact with the soil. Spreading efficiency is measured in terms of the recharge rate. Spreading methods may include flooding, basin, ditch or furrow recharge, natural channel and IRRIGATION.

Water supply sources. Water supplies for domestic, industrial and agricultural purposes are derived essentially from underground (ground water) and from surface water sources. Underground sources include supplies from wells, shafts (large-diameter wells, which in the past were normally hand dug), headings (lateral underground conduits driven out from large-diameter wells) and drifts or adits (driven at an angle from the surface into the ZONE OF SATURATION). Surface sources are mainly direct river intakes and river-impounding schemes (dams).

Water supply works sited at a spring discharge point are intermediate in definition between underground and overground sources. Another intermediate form of source is where a well is constructed to derive the greater proportion of its supply from a river, either by being sunk in close proximity to, or by driving headings underneath or parallel to, the river (Figure 17). This type of well is often referred to as a COLLECTOR WELL or Ranney well and the process of obtaining water, from the shallow unconsolidated aquifer in which the well is sited, is termed INDUCED RECHARGE.

Water table. The upper surface of the ZONE OF SATURATION (Figure 7), sometimes referred to as the phreatic surface. Everywhere below the water table ground water fills all the VOIDS and INTERSTICES. The water table thus separates the ground water zone from the CAPILLARY FRINGE in UNCONFINED AQUIFER conditions.

The water table can be found by noting water levels in a series of wells sunk into the unconfined aquifer. It can then be described on a map by ground water contours in the same way as surface topography, and is, in fact, simply a subdued reflection of the surface topography.

Because of the association between the water table and unconfined aquifer conditions, atmospheric pressure obtains along this phreatic surface. The water table thus constitutes a free surface along which pressures are uniform. It can fluctuate in elevation with recharge to and discharge from the aquifer. A rise in the water table after a period of precipitation is delayed by the time taken for the water to travel from the ground surface to the saturation zone and must take into account any deficits in soil moisture that may be present. In Britain, although the greater quantity of rain commonly infiltrates between

October and March, when evaporation is low, the maximum height of the water table may not be established until March or later.

The size of water table fluctuations in unconfined aquifers depends upon the specific yield (◊EFFECTIVE POROSITY) and permeability of the aquifer and the volume of effective INFILTRATION. For an equivalent amount of infiltration reaching the zone of saturation the rise of the water table in a highly permeable sandstone, e.g. Trias sandstone in Britain, may be no more than 1·5 metres, while in a LIMESTONE AQUIFER, e.g. the chalk, it may commonly be in excess of 8 metres. This is because SPECIFIC RETENTION is high in the small voids of the chalk whereas specific yield is high in the larger interstices of the sandstone. Water table fluctuations may also be due to both tidal and barometric effects on aquifers, but these effects are best shown by fluctuations of the PIEZOMETRIC SURFACE in CONFINED AQUIFERS.

Weathering. The chemical and mechanical breakdown of rocks and minerals under the action of atmospheric agencies.

Well. A term used for any hole in the ground for water supply (or in the oil industry for oil and/or gas), independent of depth and diameter. The well may be a dug well or shaft, drilled well, etc. The impression that a well and a borehole are quite distinct is an unreal concept; for example, drilled wells may be up to two metres in diameter.

Wells used to supply ground water for domestic, industrial or agricultural purposes are called abstraction or PRODUCTION WELLS. Occasionally they are simply referred to as pumping or discharging wells. Other wells used to monitor the DRAWDOWN characteristics around an abstraction well which are due to pumping in that well are termed OBSERVATION WELLS.

After a deep water well has been drilled, it should be completed and developed for optimum yield and tested before installing the submersible pump used for long-term supplies. For a sustained-yield performance, the well should be sealed against the entrance of surface contaminants and given periodic maintenance. Some wells contain horizontal connections (headings) or lateral tubes, including COLLECTOR WELLS and INFILTRATION GALLERIES, and are constructed in areas where special ground water conditions obtain.

Initial development of an aquifer often includes the drilling of a test well to determine the depth to ground water, i.e. depth to the ZONE OF SATURATION, the quality of the water and the thickness of the aquifer. These test holes are normally narrow in diameter and inexpensive compared with a final abstraction well. If the test hole appears suitable as a site for ground water pumping, it can be enlarged – reamed out – to a greater diameter and used as an abstraction well.

Completion of the production well must provide for the ready entrance of ground water into the well with minimum resistance in and around the CASING. Completion of the construction frequently includes the use of various well screens and GRAVEL PACKS, the type used depending on the nature of the aquifer being developed.

Occasionally wells serve other purposes such as subsurface exploration and observation, ARTIFICIAL RECHARGE of ground water, and disposal of sewage or industrial wastes. Great care has to be exercised, however, regarding the latter usage of old, abandoned wells, so that pollution of the ground water in the aquifer does not occur.

Whiteware clays. Whitewares are manufactured from a blend of CHINA CLAY or BALL CLAY, and flint or chert, with the addition of feldspar which may be replaced by talc or nepheline syenite. Whiteware clays should contain only a very small proportion of iron oxides and organic material.

Willemite. ◊ZINC.

Wilting point. That soil moisture content at which permanent wilting of plants occurs (also called wilting coefficient). Experimental work has shown that this is not a unique value, but rather depends upon the plant, the climate, the root system and the volume of soil tested.

Witherite. ◊BARIUM MINERALS.

Wolframite. ◊TUNGSTEN.

Wollastonite. This calcium metasilicate mineral ($CaSiO_3$) characteristically occurs in association with garnet, calcite and diopside in thermally metamorphosed sandy limestones. It is rarely sufficiently concentrated to be worked. A well-known deposit at Willsboro in New York State occurs in metamorphosed Precambrian limestones. Wollastonite is used in the ceramics industry; it has also been employed as an extender in paint (◊MINERAL FILLERS), as GRANULES and as a REFRACTORY material. It is potential raw material for the making of mineral wool (◊MINERAL AND ROCK WOOL). Production in 1971, in tonnes, was: USA, 40000; Finland, 5549; Mexico, 2716.

X

Xerophytes. Desert plants which are adjusted to an extreme economy of water. These plants have a shallow but widespread root system to maximize their moisture-tapping capacity.

Y

Yttrium (Y). An element which occurs closely associated with the RARE EARTH ELEMENTS. Its compounds are used in the manufacture of cathode ray tubes for colour television, in lasers and other electronic applications. It is obtained as a by-product of rare earth oxide production. World production (1968) was of the order of 230 tonnes.

Z

Zinc (Zn). A metal used as a constituent of alloys by man since pre-historic times. It has a crustal abundance of 70 ppm. Today its principal uses are as a protective coating (galvanizing) for steel and in die-casted alloys. Zinc alloys are extensively used in the construction of motor cars and household appliances. Brass is a zinc alloy widely used in household plumbing and heating installations; other uses of zinc are diverse. The annual consumption of zinc is a good barometer of the state of the economy in the countries of the Western world.

The principal ore of zinc is the sulphide (ZnS), sphalerite or 'blende'. Part of the zinc is usually replaced by iron, and cadmium is a common impurity. Sphalerite weathers readily and is found in association with a variety of compounds including zincite (ZnO), smithsonite ($ZnCO_3$) and the silicates, hemimorphite and willemite. FRANKLINITE (a complex iron–zinc–manganese oxide of the SPINEL group) occurs in the unique, but commercially important, deposits of New Jersey, USA. Small amounts of GERMANIUM, GALLIUM, INDIUM and THALLIUM may be recovered as by-products during the processing of zinc ores.

Most commercial deposits are cavity fillings or replacements, especially in limestones, from solutions of probable magmatic origin. They are often associated with lead (galena), iron (pyrites) and copper sulphides.

Commercial deposits are widely distributed; over 50 countries produce some zinc ore but Canada (1·2 million tonnes) supplied a quarter of the world total (4·8 million tonnes) in 1974. After Canada, the USSR, USA, Australia and Peru were the main producers.

Zincite. ▷ZINC.

Zircon. ▷ZIRCONIUM.

Zirconium (Zr). Zirconium metal is used in the construction of nuclear reactors, in corrosion-resistant alloys for the chemical industry and in the manufacture of photo flashbulbs. Most zirconium, however, is used in the form of the source mineral, zircon, particularly as a constituent of FOUNDRY SANDS and other REFRACTORIES. The oxide is used as a pigment in ceramics and pottery.

Zircon is produced almost entirely from heavy mineral sands (PLACER DEPOSITS) where it is associated with RARE EARTH and TITANIUM minerals. World production of zircon sand in 1973 was of the order of 500000 tonnes. Australia, the USA and USSR were the major producers.

Zone of aeration. ▷AERATION ZONE.

Zone of oxidation, oxidized enrichment. ▷SECONDARY ENRICHMENT.

Zone of saturation. That part of an aquifer where ground water fills all the INTERSTICES. Thus the zone of saturation exists everywhere beneath the WATER TABLE or PIEZOMETRIC SURFACE (Figure 7). In this zone POROSITY is a direct measure of the water contained per unit volume but not all the water can be removed by pumping from this zone, as a certain amount is held back in the rock fabric by molecular and surface-tension forces. The zone of saturation may include both permeable and impermeable rock materials.

Whereas movement of subsurface water in the AERATION ZONE is essentially downward in direction, the movement of ground water (saturated flow) in the zone of saturation is governed by the hydrostatic pressure and may be in any direction, with a lateral component usually being the more important.

This saturated flow or PERCOLATION in the saturated zone takes place from a high inflow head to a lower outflow head, that is in the direction of the HYDRAULIC GRADIENT. (▷UNSATURATED ZONE.)

Guide to Further Reading and Sources of Information

compiled by C. J. Spittal and Jennifer Scherr

General Texts

A. M. Bateman, *Economic Mineral Deposits*, Wiley, 2nd edn, 1950.

R. L. Bates, *Geology of the Industrial Rocks and Minerals*, Dover, rev. edn, 1969.

L. Bertin, *The New Larousse Encyclopaedia of the Earth*, Hamlyn, rev. edn, 1972.

R. A. Deju and others, *The Environment and its Resources*, Gordon and Breach, 1972.

R. A. Deju (ed.), *Extraction of Minerals and Energy: Today's Dilemmas . . .*, Ann Arbor Science, 1974.

W. G. Ernst, *Earth Materials*, Prentice Hall, 1969.

A. J. Fagan, *The Earth Environment*, Prentice Hall, 1974.

R. W. Fairbridge (ed.), *Encyclopedia of Earth Sciences Series*, Van Nostrand, Reinhold. Vol. I: *Encyclopedia of Oceanography*, 1966; vol. II: *Encyclopedia of Atmospheric Sciences and Astrogeology*, 1967; vol. III: *Encyclopedia of Geomorphology*, 1968; vol. IVA: *Encyclopedia of Geochemistry and Environmental Sciences*, 1972. The series is continuing and will cover mineralogy and economic geology, among other subjects.

P. T. Flawn, *Environmental Geology: Conservation, Land-Use Planning and Resource Management*, Harper & Row, 1970.

P. T. Flawn, *Mineral Resources: Geology, Engineering, Economics, Politics, Law*, Wiley, 1971.

F. E. Frith (ed.), *Encyclopedia of Marine Resources*, Van Nostrand, 1969.

G. Holister and A. Porteous, *The Environment: A Dictionary of the World around Us*, Arrow Books, 1976.

C. A. Lamey, *Metallic and Industrial Mineral Deposits*, McGraw-Hill, 1966.

J. F. McDivitt and G. Manners, *Minerals and Men . . .*, Johns Hopkins, rev. and enl. edn, 1974.

H. W. Menard, *Geology, Resources, and Society: an Introduction to Earth Science*, W. H. Freeman, 1974.

J. L. Mero, *The Mineral Resources of the Sea*, Elsevier, 1965.

J. E. Metcalfe, *British Mining Fields*, Institution of Mining and Metallurgy, 1969.

Mineral Resources Consultative Committee, *Mineral Dossiers*, HMSO, published annually.

National Academy of Sciences: National Research Council, *The Earth and Human Affairs*, Canfield Press, 1972.

National Academy of Sciences: National Research Council, Committee on Resources and Man, *Resources and Man; a Study and Recommendations*, W. H. Freeman, 1969.

Open University, *Science: A Second Level Course*, S26, *The Earth's Physical Resources*. Block 1: Resources and systems; Block 2: Energy resources; Block 3: Mineral deposits; Block 4: Constructional and other bulk materials; Block 5: Water resources; Block 6: Implications: limits for growth?

W. N. Peach and J. N. Constantin, *Zimmerman's World Resources and Industries*, Harper & Row, 3rd edn, 1972.

R. H. S. Robertson, *Mineral Use Guide or Robertson's Spiders' Webs*, Cleaver Hume, 1971.

B. J. Skinner, *Earth Resources*, Prentice Hall, 1969.

K. K. Turekian, *Oceans*, Prentice Hall, 1968.

W. Van Royen and others, *The Mineral Resources of the World*, Prentice Hall, 1952.

Metallic Ores

A. M. Bateman, *Economic Mineral Deposits*, Wiley, 2nd edn, 1950.

R. L. Bates, *Geology of the Industrial Rocks and Minerals*, Dover, rev. edn, 1969.

J. Blunden, *The Mineral Resources of Britain: a Study in Exploitation and Planning*, Hutchinson, 1975.

R. A. Deju (ed.), *Extraction of Minerals and Energy: Today's Dilemmas . . .*, Ann Arbor Science, 1974.

K. C. Dunham, 'Geological settings of useful minerals in Britain', *Proceedings of the Royal Society of London*, series A, vol. 339 (1974), pp. 273–88.

P. T. Flawn, *Mineral Resources: Geology, Engineering, Economics, Politics, Law*, Wiley, 1971.

J. L. Gilleson (ed.), *Industrial Minerals and Rocks*, American Institute of Mining, Metallurgical and Petroleum Engineers, 3rd edn, 1960.

Institution of Mining and Metallurgy, *Opencast Mining, Quarrying and Alluvial Mining . . .*, Institution of Mining and Metallurgy, 1965.

C. A. Lamey, *Metallic and Industrial Mineral Deposits*, McGraw-Hill, 1966.

J. F. McDivitt and G. Manners, *Minerals and Men . . .*, Johns Hopkins, rev. and enl. edn, 1974.

J. L. Mero, *The Mineral Resources of the Sea*, Elsevier, 1965.

J. E. Metcalfe, *British Mining Fields*, Institution of Mining and Metallurgy, 1969.

Mineral Resources Consultative Committee, *Mineral Dossiers*, HMSO, published annually.

B. C. Netschert and H. H. Landsberg, *The Future Supply of the Major Metals*, Resources for the Future, Washington, 1961.

C. F. Park and R. A. Macdiarmid, *Ore Deposits*, Freeman, 2nd edn, 1970.

R. H. S. Robertson, *Mineral Use Guide or Robertson's Spiders' Webs*, Cleaver Hume, 1971.

W. Van Royen and others, *The Mineral Resources of the World*, Prentice Hall, 1952.

Building Materials

A. Clifton-Taylor, *The Pattern of English Building*, Faber, 1972.

M. H. Grant, *The Marbles and Granites of the World*, J. B. Shears, 1955.

P. S. Keeling, *The Geology and Mineralogy of Brick Clays*, Brick Development Association, 1963.

H. O'Neill, *Stone for Building*, Heinemann, 1965.

B. C. G. Shore, *The Stones of Britain*, Leonard Hill, 1957.

Chemical Deposits

J. L. Gillson and others (eds.), *Industrial Minerals and Rocks (Non-Metallics Other than Fuels)*, Amer. Inst. Mining, Metallurgical and Petroleum Engineers, 3rd edn, 1960.

R. B. Ladoo and V. M. Myers, *Non-Metallic Minerals*, McGraw-Hill, 2nd edn, 1951.

Energy, Fossil Fuels, Radioactive or Nuclear Power

E. R. Berman, *Geothermal Energy*, Noyes Data Corporation, 1975.

Committee for Environmental Conservation and others, *Energy and the Environment*, Royal Society of Arts, 1974.

W. Francis, *Coal: its Formation and Composition*, Arnold, 2nd edn, 1961.

H. E. Goeller and others, *World Energy Conference Survey of Energy Resources*, US National Committee of the World Energy Conference, New York, 1974. (Subsequent editions to be issued at six-yearly intervals.)

G. D. Hobson and E. N. Tiratsoo, *Introduction to Petroleum Geology*, Scientific Press, 1975.

K. A. D. Inglis (ed.), *Energy: From Surplus to Scarcity? . . .*, Applied Science Publishers, 1974.

International Petroleum Encyclopedia, Petroleum Publishing Co., Tulsa, Oklahoma, 1970.

D. E. Kash and others, *Energy under the Oceans: a Technology Assessment of Outer Continental Shelf Oil and Gas Operations*, Bailey Bros. & Swinfen, 1974.

A. I. Levorsen, *Geology of Petroleum*, Freeman, 2nd edn, 1967.

G. Manners, *The Geography of Energy*, Hutchinson, 1971.

J. B. Marion, *Energy in Perspective*, Academic Press, 1974.

National Economic Development Office, *Energy Conservation in the United Kingdom: Achievements, Aims and Options*, HMSO, London, 1974.

Organization for Economic Co-operation and Development, *Energy Prospects to 1985: an Assessment of Long-Term Energy Developments and Related Policies*, 2 vols, OECD, 1975.

L. A. Redman, *Nuclear Energy*, Oxford University Press, 1963.

D. Z. Robinson and others (eds.), *Nuclear Energy Today and Tomorrow*, Heinemann, 1971.

L. C. Ruedisili and M. W. Firebaugh (eds.), *Perspectives on Energy: Issues, Ideas and Environmental Dilemmas*, Oxford University Press, 1975.

Scientific American, *Energy and Power*, W. H. Freeman, 1971.

M. W. Thring and R. J. Crookes (eds.), *Energy and Humanity*, Peter Peregrinus, Stevenage, 1974.

E. N. Tiratsoo, *Natural Gas: a Study*, Scientific Press, 1972.

E. N. Tiratsoo, *Oil-fields of the World*, Scientific Press, 1973.

I. A. Williamson, *Coal Mining Geology*, Oxford University Press, 1967.

R. Wilson and W. J. Jones, *Energy, Ecology and the Environment*, Academic Press, 1974.

Water

P. Briggs, *Water, the Vital Essence*, Harper & Row, 1967.

S. N. Davis and R. J. N. De Wiest, *Hydrology*, Wiley, 1966.

D. M. Gray (ed.), *Handbook on the Principles of Hydrology, with Special Emphasis Directed to Canadian Conditions . . .*, Canadian Nat. Committee for the International Hydrological Decade, 1970.

Ground Water and Wells, E. E. Johnson Incorporated, St Paul, Minnesota, 1966.

Ground Water Yearbook, HMSO, published annually.

B. Horsfield and P. B. Stone, *The Great Ocean Business*, Hodder & Stoughton, 1972.

L. B. Leopold, *Water: a Primer*, W. H. Freeman, 1974.

R. L. Nace, *Water and Man: a World View*, UNESCO, 1969.

Scientific American, *The Ocean*, W. H. Freeman, 1969.

Scientific American, *Oceanography*, W. H. Freeman, 1971.

K. Smith, *Water in Britain: a Study in Applied Hydrology and Resource Geography*, Macmillan, 1972.

Surface Water Yearbook of Great Britain, HMSO, published annually.

D. K. Todd, *Ground Water Hydrology*, Wiley, 1959.

D. K. Todd (ed.), *The Water Encyclopedia: a Compendium of Useful Information on Water Resources*, Water Information Center, Port Washington, N.Y., 1970.

W. C. Walton, *Groundwater Resource Evaluation*, McGraw-Hill, 1970.

Water Resources Board, *Water Resources in England and Wales*, 2 vols, HMSO, London, 1973. (Water Resources Board Publications, 22 and 23.)

J. Williams, *Oceanography: an Introduction to the Marine Sciences*, Little & Brown, 1962.

C. O. Wisler and E. F. Brater, *Hydrology*, Chapman & Hall, 1959.

Abstracting, Indexing and Current Awareness Services

British Geological Literature, Nos. 1–5, Corindon Press, London, 1964–8; N.S. Brown's Geological Information Service, London.

Current Contents: Agriculture, Biology and Environmental Sciences, Institute for Scientific Information, Philadelphia, Pa.

Current Contents: Engineering, Technology and Applied Sciences, Institute for Scientific Information, Philadelphia, Pa.

Current Contents: Physical and Chemical Sciences, Institute for Scientific Information, Philadelphia, Pa.

Deep Sea Research and Oceanographic Abstracts, Pergamon Press.

Earth Resources: a Continuing Bibliography (NASA SP–7041), NASA, Washington.

Geotitles Weekly, Geosystems, London.

Marine Science Contents Tables, FAO, Rome.

Mineralogical Abstracts, Mineralogical Society, London.

Petroleum Abstracts, University of Tulsa, Tulsa, Okla.

Water Resources Abstracts, American Water Resources Association, Urbana, Illinois.

Dictionaries and Directories

M. Gary and others (eds.), *Glossary of Geology*, American Geological Institute, 1972.

P. Hepple, *A Glossary of Petroleum Terms*, Institute of Petroleum, 4th edn, 1967.

S. R. Kaplan (ed.), *A Guide to Information Sources in Mining, Minerals and Geosciences*, Interscience, 1965.

Mine and Quarry (formerly *Mining and Minerals Engineering*; formerly *Mine and Quarry Engineering*), Ashire Publishing Ltd, London.

Mining International Yearbook, Financial Times, London.

A. Nelson and K. D. Nelson, *Dictionary of Applied Geology*, Butterworth, 1967.

A. Nelson, *Dictionary of Mining*, Newnes-Butterworth, 2nd edn, 1973.

E. J. Pryor, *Dictionary of Mineral Technology*, Mining Pubns, 1963.

Skinner's Oil and Petroleum Yearbook, Financial Times, London.

P. W. Thrush, *A Dictionary of Mining, Mineral and Related Terms*, GPO, Washington, 1968.

D. G. A. Whitten and J. R. V. Brooks (eds.), *A Dictionary of Geology*, Penguin, 1972.

H. Wöhlbier and others (eds.), *Worldwide Directory of Mineral Industries Education and Research*, Gulf Publishing Co., Houston, Texas, 1968.

Journals

Bulletin of the American Association of Petroleum Geologists, AAPG, Tulsa, Okla.

Bulletin: United States Geological Survey, GPO, Washington.

Economic Geology, Economic Geology Publishing Co., Houghton, Michigan.

Geotimes, American Geological Institute, Washington.

Groundwater, Water Well Journal Publishing Co., Columbus, Ohio.

Hydrological Sciences Bulletin, Institute of Hydrology, Wallingford, Berks.

Minerals Yearbook, GPO, Washington.

Mining Annual Review (annual supplement to *Mining Magazine* and *Mining Journal* published each June), Mining Journal, London.

Mining Journal, Mining Journal, London.

Mining Magazine, Mining Journal, London.

Nature and Resources, Newsletter about scientific research on environment, resources and conservation of nature, UNESCO.

Professional Papers: United States Geological Survey, GPO, Washington.

Special Reports on Mineral Resources: Economic Memoirs, Geological Survey of Great Britain, HMSO.

Transactions: American Institute of Mining, Metallurgical and Petroleum Engineers, AIMMPE, New York.

United Kingdom Mineral Statistics, HMSO.

Water Resources Research, American Geophysical Union, Washington.

53899

Other Books in the Series

Arrow dictionaries have been written and designed to open up expanding specialist areas of knowledge to the non-specialists, both students and general readers.

Other books currently in the series are:

Professor Geoffrey Holister and Dr Andrew Porteous

THE ENVIRONMENT

a dictionary of the world around us

Tim Congdon and Douglas McWilliams

BASIC ECONOMICS

a dictionary of terms, concepts and ideas

Dr Anthony Hyman

COMPUTING

a dictionary of terms, concepts and ideas

Professor Geoffrey Holister and Dr Andrew Porteous

THE ENVIRONMENT
a dictionary of the world around us

Saving our beleaguered environment – the most pressing problem facing man in the second half of the twentieth century.

As concern for the environment has grown, so has a new academic discipline, embracing such previously disparate areas of knowledge as ecology, economics, physics and sociology.

This dictionary provides both the student and the general reader with a working knowledge of environmental studies – terminology, components and technology.

The authors, making no apologies for presenting a committed view of their subject, have written a reference book that will both stimulate and inform all concerned with the world's environmental crisis.

Professor Geoffrey Holister is Professor of Engineering Science at the Open University.

Dr Andrew Porteous is Reader in Engineering Mechanics at the Open University and General Editor of the OU course, Environmental Control and Public Health.